CANADIAN FAILURES

CANADIAN FAILURES

Stories of Building Toward Success

Alex Benay

DUNDURN
TORONTO

Cover image: istockphoto.com/muzon
Printer: Webcom

Library and Archives Canada Cataloguing in Publication

Benay, Alex, author
    Canadian failures : stories of building toward success / Alex Benay.

Includes bibliographical references.
Issued in print and electronic formats.
ISBN 978-1-4597-4043-3 (softcover).--ISBN 978-1-4597-4044-0 (PDF).--
ISBN 978-1-4597-4045-7 (EPUB)

    1. Success--Canada. 2. Successful people--Canada. 3. Failure (Psychology)--Canada. I. Title.

BF637.S8B44 2017   650.1   C2017-903665-3
                                                                  C2017-903666-1

1 2 3 4 5   21 20 19 18 17

 Conseil des Arts du Canada   **Canada Council for the Arts** Canada  **ONTARIO ARTS COUNCIL CONSEIL DES ARTS DE L'ONTARIO** an Ontario government agency un organisme du gouvernement de l'Ontario

We acknowledge the support of the **Canada Council for the Arts**, which last year invested $153 million to bring the arts to Canadians throughout the country, and the **Ontario Arts Council** for our publishing program. We also acknowledge the financial support of the **Government of Ontario**, through the **Ontario Book Publishing Tax Credit** and the **Ontario Media Development Corporation**, and the **Government of Canada**.

Nous remercions le **Conseil des arts du Canada** de son soutien. L'an dernier, le Conseil a investi 153 millions de dollars pour mettre de l'art dans la vie des Canadiennes et des Canadiens de tout le pays.

Care has been taken to trace the ownership of copyright material used in this book. The author and the publisher welcome any information enabling them to rectify any references or credits in subsequent editions.
    — *J. Kirk Howard, President*

The publisher is not responsible for websites or their content unless they are owned by the publisher.

Printed and bound in Canada.

VISIT US AT

 dundurn.com | @dundurnpress | dundurnpress | dundurnpress

Dundurn
3 Church Street, Suite 500
Toronto, Ontario, Canada
M5E 1M2

CONTENTS

INTRODUCTION

This is a book about Canadian failures.

This book will make you uncomfortable — well, at least a little uncomfortable (we *are* Canadians, after all). If discussing uncomfortable topics is not something that appeals to you, you may want to put this book down — now.

○ ○ ○

I'm happy you are still with us. Let's get started.

Successes and their subsequent histories define people and nations. One could argue that these stories of human ingenuity have helped to define the psyche of the entire human race for millennia. As humans we have, since the earliest of times, transformed the environment around us, innovated, and shaped the landscape to suit our needs. We are a dominant species that has thrived on a history of striving for success. Technological change has permeated our development, from the use of fire to the wheel, the printing press and, more recently, mass communication technologies, sparking a digital revolution that has left no facet of our global society untouched. In this context, humans have also learned to deal with failures and setbacks. Shuttles destined for faraway space travel have exploded, and while we have eradicated some diseases, others have spread. We live with more stories of failure than of success, yet we often only speak of success.

Canada is not immune to this. We are taught from a young age that our great country was built on the back of the railroad and the telegraph, how we are global leaders in the agri-food industry, how we discovered insulin, and a multitude of other successes. These stories of success are what make us Canadians, or so we are told from the moment we enter school. When I was the head of a national heritage institution, one responsible for the safekeeping of stories of our country's ingenuity, I found myself immersed in artifacts and archives of great Canadian successes, such as actual steam engines from the twentieth century and blueprints for new automotive vehicles of the 1960s, along with a multitude of other "world firsts." Museums such as the one I had the pleasure of serving are littered with stories of success. I disseminated our great nation's stories of success and ensured their safekeeping — that was my job. This role I played helps to contribute to the overall fabric of our national identity. It is a valuable role. However, these great success stories only represent a small segment of our national character. A nation that prides itself only on its success is not prepared to face the challenges of the future. A nation should face its failures, accept them, and in certain cases even celebrate them, in an effort to become greater than it is.

Indeed, a nation is equally defined by its failures and its successes. It might be said that a nation is even *more* defined by failure than by success. However, a dialogue on failures is not in any educational curriculum. Our heritage continues to be celebrated as a series of successes. Rarely is our identity positioned by our failures as an inclusive whole — a timeline of human ingenuity, coupled with the realities of failure. Failure, while it is discussed in corporate contexts, is still not celebrated openly as a means to a greater end. In a world that is often described as entering a new "disruptive" phase of existence, complete with disruptive technologies, disruptive economies, and so forth, and at a time when the innovation economies of the world are seen as a critical next step in our human evolution, not failing often and fast seems counterintuitive to any innovation agenda. Yet for innovation and disruption to stop being mere buzzwords, and if we are to truly embrace failure as a means to greater achievement, we must first start by acknowledging failure, and enter into a dialogue about our short-comings. We have to be brutally honest with ourselves.

We speak of a historic rail project, with only minimal reflection on our nation's abuse of minority populations in order to reach the great "last spike." We do not speak openly across the entire nation about the atrocities committed against our First Nations. We do not engage with the environmental issues resulting from our consumerism. We do not speak to the failures of our science system, which in many ways is too Canadian, too polite, yet belligerent toward this country's women. We do not speak of the failures of our Canadian enterprises in achieving true global status: this lack of large Canadian enterprise creates a fundamental gap in our national DNA, for the United States' identity is as much defined by Ford and Apple as it is by Abraham Lincoln and Mark Twain.

To make things worse, we perpetuate this situation through our literature. Unfortunately, when most authors set themselves a goal to review and discuss the national identity, the typical path is to proceed to illustrate accomplishments: how great the country is as a result of its economic record on the world scene, or perhaps its greatest inventions, or its multitude of amazing corporations — think Germany or Japan. In other instances, books list a country's cultural impact and the influences of its authors, artists, and scholars — think France. The challenge with these points of view is that they represent only a small sliver of the national identity dialogue. Approaching a debate on national identity with these blinders on is wrong. It is wrong because speaking only of accomplishments is taking the easy road. Museums, archives, and government institutions excel at collecting objects, documents, and other such constructs of success, because it is easier to do this than to speak of, and collect, failure.

Not openly engaging in a dialogue about our failures is *the* critical mistake we can make as a nation, because a dialogue on failures inevitably leads to a dialogue on national identity: failures define us as much as successes; they shape our national DNA, our culture, and our creative spirit. In the following chapters, individuals who have both failed and succeeded will address our reluctance as a nation to speak about Canadian failures and how, through time, failures have helped to shape our national identity. They will address our uniqueness as a nation by addressing our failures, not our accomplishments.

At this juncture, we should ask: "Why conduct this exercise in the first place?" The answer is relatively simple: Because not every citizen identifies with successes. Because different generations, ethnicities, and genders see success differently; what was a success for one person may have been a complete failure for another. Because not speaking of our failures does not help us to grow as a nation — in fact, speaking only of successes makes us weaker as a nation. The participants in this project believe in Canada, so we speak about our failures to make this country even greater.

Why should a museum spearhead such a project? Because, as the national institution responsible for developing a science and innovation culture across the country, Ingenium — Canada's Museums of Science and Innovation (formerly Canada Science and Technology Museums Corporation) represents everything that is traditional about celebrating Canadian successes. It has perpetuated a certain type of myth of Canadian greatness and has not spoken of how our failures have equally contributed to our national identity. Two Torontonians invented a working version of the incandescent lightbulb and sold their patent to Edison. We all know what he did with the lightbulb. There also is the Avro Arrow, Nortel, and many other examples that have had profound impacts on our national identity. Yet we do not speak of the impacts of these failures on what it means to be Canadian at the same levels of admiration with which we speak of insulin, Alexander Graham

Bell, or even Elon Musk. We latch onto success and fail our nation by not engaging in a dialogue on failure.

As a result of this reluctance, the following book is a Canadian first. It is an attempt to bring together a diverse set of authors from different generations and ethnic backgrounds to talk candidly about our Canadian failures. Professors, business women, government officials, and others provide their vantage points on important failures and how these have shaped our identity as a nation. These authors are the courageous ones who have chosen to speak about their personal stories, their feelings about failure, and what Canada could be doing better, all in the hope of sparking a national dialogue. You should not agree with them at every step of the way; that is contrary to the purpose of this process. But you should not approach this book with the mindset that Canada is perfect, because it is not. We are no less Canadian because we speak of failure — in fact, I would argue we are prouder of our country because we do so.

Furthermore, as you read through this book, please remember that writing it was not an easy task. What I have asked these great Canadians to do is not easy. I have asked them to put their failures out there, out into the world for all to witness, in the hope of creating a stronger nation. At the very least, I hope this book creates a platform for our heritage to include our failure on an equal footing with our success.

This is a book about Canadian failures because we are proud Canadians.

— Alex Benay

Chief Information Officer, Government of Canada; former President and CEO of the Canada Science and Technology Museums Corporation (now called Ingenium – Canada's Museums of Science and Innovation)

AMIEE CHAN

President and CEO, Norsat International Inc.

Tuning into Success at Norsat International

I have never met Amiee in person. I know this may sound weird, but her reputation precedes her, and we have spoken on the phone a few times. She left her mark on the Canada Science and Technology Museums Corporation (now Ingenium) as a member of the Board of Trustees. Years after her departure, her presence is still felt around the institution, and she is sorely missed. She has been the CEO at Norsat International, a global telecommunication and satellite company, since 2006. She also holds an M.B.A. from Simon Fraser University and a Ph.D. in Satellite Communication from the University of British Columbia. As a woman in technology, she is a role model; she is also a community leader and an entrepreneur at heart. I could not think of anyone more suited to speak to entrepreneurialism and women's issues in our great country than Amiee. I just wish I'd had the chance to work with her directly.

— Alex Benay

The story of Norsat International Inc. from the late 1990s to the 2000s is one of mini-booms and mini-busts, successes and failures, false starts, and rebooting a company. At its peak in 2001, Norsat stock was trading at $46 a share, and $460 after consolidation. Amazingly, in British Columbia — a province of giant mining and forestry natural resource companies — Norsat was the fifth largest company by market capitalization. In its heyday, when Norsat was bidding on big contracts, it had about four hundred and fifty employees. At a low point, the number dwindled to about forty-five employees.

At its beginning, Norsat was a really, really small company, but there were huge aspirations to grow quickly. Before long the company was making under $20 million per annum in revenue, but was bidding on $100 million contracts. Even if Norsat had won the contracts, I'm not sure it could have fulfilled them successfully.

Norsat had always been a leader in making products in the satellite communications field for private and public companies, broadcasters, and the military. It had always had a great product line, which was supported by excellent intellectual property and great team of engineers. But the company bet the farm trying to land big military contracts. Norsat hired high-priced executives from the United States and spent a lot of money on travel, trying to get large contracts that the company thought would get it where it wanted to go.

It also dabbled in low-margin, high-volume products, like satellite television for the Canadian home market. Norsat struggled to compete in this market, though, as more and more of these products were made in China, and electronics became cheaper. It became impossible for the company to match competitors' pricing and maintain margins.

Norsat was always trying different strategies. As a public company, it had the confidence of the markets and was able to raise money to fund the strategies as they came and went, by issuing debt and selling more shares. I have to say, it was pretty successful at raising money. Every time it got an influx of cash the spending would start again, and the company would go all-in on the next big thing.

In the 1990s and early 2000s, the tech sector was still attracting investors and investments. But shortly afterward, when the tech sector went up in flames (think Nortel), everyone was afraid to invest in high tech. The capital markets dried up and a lot of money left the sector. Nobody wanted to touch it. We couldn't go to the market to raise money.

I was part of the management team then. After Norsat's big spending and recent difficulty raising new money, the company was in dire straits. Norsat had lost a lot of money. We had trouble sustaining the company, and employees were walking around wondering if they were even going to get their next paycheque.

It was a really tough situation.

But feast or famine was the history of Norsat.

Norsat and Me

I've always had a passion for satellite communications, and as part of my Ph.D. thesis at the University of British Columbia, I worked on developing the next generation of satellite terminals. I knew that mobility was a big thing: users wanted to be sure of their connection and able to move around. The concept of a portable satellite terminal had always interested me, and I thought a flat-panel antenna was key to improving the concept.

Terminal portability was part of our studies and our investigations at UBC. The school had a contract with NASA to work on the next generation of satellite communications. That meant we got to see what the Jet Propulsion Laboratory and NASA were doing. We also collaborated with a group of universities around the world in a project called the NASA ACTS project. Its focus was the next generation of satellite communications technology, and terminals were a part of it. I concentrated on the terminal part; I wrote a lot of papers, and I did a lot of studies. I became a subject-matter expert and I talked to a lot of people. Meanwhile, Norsat was focusing on components that go into satellite terminals.

While studying for my graduate degree at UBC, I worked full time at a local high-tech company. I ended up at Norsat when it bought the division of the company where I was working. I made an arrangement with Norsat so that I could attend UBC part-time at night and also work at the company. I had the best of both worlds — combining university work that involved NASA and gaining expertise by working at Norsat.

∘ ∘ ∘

One of Norsat's main products is the electronics that go into satellite dishes. Anytime you see a satellite dish on a roof for a commercial application — such as a gas station, a car dealership, or a Walmart in a small town — there's a very good chance that the electronics came from Norsat, because we've always been the leader. But being the leader, there wasn't much more market share to take.

So Norsat was struggling to find a new market to break into. I started pushing for the company to get into the next generation of portable satellite terminals. After all, this was the subject of my thesis and what I was passionate about. My idea was a satellite terminal like those you see on the rooftops of buildings, but one that you could fit into a backpack, jump out of a helicopter with anywhere in the world, put together in five minutes, and plug your laptop into, to connect instantly to the internet.

At the time, we had a customer who had seen some of our papers on portable satellite terminals and really liked what it saw. This customer asked if we could make one for its operations. This was a good situation to find ourselves in, because Norsat didn't have a lot of money and needed a new initiative. Suddenly we had a paying customer and a great idea — it seemed as if all the stars had aligned!

∘ ∘ ∘

We started development of portable satellite terminals in consultation with the customer. The client wanted the product to do this

Amiee Chan inspecting a satellite terminal component.

and that — and all sorts of things. It had to be light as well as small. "Make it so it fits in a briefcase so a James Bond–type can take it with him and set it up quickly and communicate when he reaches the destination." The customer wanted it to be small enough and light enough be carried around easily, and to be able to integrate with a laptop. "We want it to have a writing pad. And it should have a light so it can be used in the dark." Once the basic concept and features of a portable satellite terminal had taken shape, a lot of Norsat people contributed specific ideas. Some ideas were more practical than others. We got all our engineers in the room — electrical, mechanical, software — and everybody put in their two cents.

Our first prototype was novel, but it wasn't very functional or practical. We made it small enough to fit in a briefcase, but it was too small to talk to a lot of different satellites. With its limited range it could only access certain satellites. So it was a good fit for that one customer who needed it only for lower data rates, but not necessarily for other customers who needed higher speed communications.

Its mechanics were a little bit awkward, too. It was fiddly. To find a satellite you had to tune the reception mechanically. You had to turn stiff knobs to tune the azimuth, or elevation. To be honest, it was clunky and inelegant.

But we had a first prototype that was working, and we celebrated. However, when we tried to sell it to another customer, this customer found many other faults. Still, we weren't discouraged. Norsat had just gotten another cash infusion, so we said, "Let's make the next generation of portable satellite terminal for a wider set of customers."

Not long after, Norsat again fell into some difficult times and laid me off. About eighteen months later it received another cash infusion, changed its senior management team, and brought me back as Vice President of Engineering and Operations, looking to me to lead the team developing the next generation of satellite terminals. I still had to be interviewed for the job! The new CEO said that he had heard great things about what I had done in the past, and he needed me back because the product had some serious quality issues that needed fixing.

What I haven't told you is that I was originally laid off as the result of a failure on my part. I'll come back to that later.

Accepting the Challenge

I returned to Norsat, leaving a great job with another company because the position of VP of Engineering and Operations dovetailed with one of my life goals: to turn my thesis into reality. How many people ever get to do that?

So here I was, back at the company, heading up this important project. We talked to more customers. We got feedback from our customer service people in the field. We changed a lot of materials. We made our terminal sturdier and more rugged. Our second-generation terminal became *the* product for broadcasters. Unfortunately, at that time — 2004 to 2005 — tech was in a slump, and broadcasting was in even more of a slump. Advertising on TV was down, people were switching to the internet, and cable subscriptions were dropping, too. We decided that, given the market situation, we'd turn our focus to the military market. We already had some traction there, as there was already military interest in our products. It was in the early years of the wars in Afghanistan and Iraq, and our timing looked to be spot-on.

We rejigged the portable satellite terminal to appeal to different customers. We did some simple things, like painting the terminals white for broadcasters and desert tan for the military. Broadcasters had been using satellites for broadcasting for a long time. They knew how to line up a terminal to find a satellite. But in the military, many of the users were nineteen years old, right out of school, and they didn't even know what satellite communication was. That meant that we had to go one step further and make a terminal that could automatically find the satellite on its own.

It needed to be really simple — as in "just push a button." We had to take a lot of the manual interface steps — connecting a cable, turning on a switch — and convert them all into software, because users (a Red Cross nurse, military personnel, even a TV correspondent) just wanted to click a mouse. They didn't even want to flip a switch. Our users wanted to work off a screen, as they did with a computer. A screen also meant we could make the interface function in multiple languages. That was a great advantage, because there were so many other countries going into the war zone who could also be our customers. So we made the terminal available in seventeen languages, and we made it really easy to use. Our goal was that if you know how to use a mouse you should be able to operate a satellite terminal: the workflow was a lot easier that way. We made it even more rugged so it could fall and not break. We put it in a waterproof case so it didn't matter if it landed in a swamp. We made the terminal a lot more durable for the war zone.

Despite the promising future of the portable satellite terminal, Norsat again ran out of money. Once more, the company was in dire straits. Our directors asked, "How do we clean up this mess?"

They turned to me and asked if I would be willing to take a shot as CEO.

My immediate reaction was to think, "Oh my goodness, look at Norsat's situation!" We had huge liabilities. We had suppliers to pay. We had employees to pay. We had customers looking for their products. We had shareholders looking at us anxiously. I considered my personal situation. I had two young children who were depending on me. I knew if I took the job it was going to be a lot of travel and a lot of personal sacrifices. Then I looked at the employees around me: they would all be unemployed if the company wound up. We had obligations everywhere, but I just couldn't say, "Let's close up shop." There was so much riding on Norsat's survival that I couldn't say no to giving it a shot. I took up the challenge.

I called a town hall meeting with all the employees to tell them that the board had asked me to step in as CEO, to change the direction of the company and turn it away from how it had been run for so long. We wanted Norsat to become self-sustaining. We wanted to be able to stand on our own two feet. And the employees were all behind me. They were incredibly supportive; I think part of it was because my path to becoming CEO had started at the bottom. I had been working with them since I was a co-op student, and we were like a family. I wasn't parachuted in: many of the people who were now my employees had once been my supervisors. They knew what I was like — that I was an engineer at heart and I could be trusted to tell it like it is. I'm very transparent, and I think everyone appreciated that.

To make the company sustainable, we had to look at everything. When I took up the CEO's role, one of the first, and hardest, things I had to do was to lay off some employees. I knew we couldn't sustain the company financially unless we did. We went down to forty-five people. At the time, we were working on the third floor of a very posh facility in Burnaby. It wasn't very functional, but the boardroom sure was nicely done — glass and chrome everywhere! We decided to go for functionality and practicality. So we moved to Richmond to be closer to the land transportation routes, and closer to the airports for shipping our products.

Our employees pulled together and said, "Just tell us what we have to do. We're behind you all the way" — and they were amazing! We only had one full-time shipper-receiver, so everyone worked together to get the product out the door. We couldn't afford someone dedicated to only answering the phone, so we combined that role with human resources. We all wore multiple hats to bring the company back on its feet. Our team did an amazing job.

We brought in advisers such as an executive coach, who was a former electronics corporation chairman and later a Norsat board member, to facilitate my transition to CEO. He gave me advice on important decisions like fine-tuning the management team and bringing in my own team members where necessary. The big-picture advice he gave me was this: Do it quickly and do it soon — all within the first week!

Changing Our Customer Acquisition Strategy

Norsat's former strategy was to go after very large contracts. As a small company, we weren't going to succeed at trying to land the large-volume, low-margin contracts because we didn't have the mass-production abilities. We decided to change strategies. We chose to customize our products for what we called low-volume, high-touch customers. These smaller customers were being left by the wayside, and we saw an opportunity. They weren't getting attention like the big customers, but they had a real need for product, and our competitors' off-the-shelf stuff wasn't working for them. Their products needed customization. We knew that our ability to design quickly and design well was our core competency, so we concentrated on that.

We started focusing on the little guys and left the big guys alone. We also decided to pursue military clients around the world in addition to the United States. The U.S. has the biggest budget, but there are a lot of smaller countries that can't afford to buy products from giant General Dynamics. I told my sales guys, "If the customer is looking for five thousand units, send them to our competition. If they're looking for five or ten, have them come to us, and we'll do whatever they need. If they want it the colour of desert sand or if they want it to run the terminal off a car battery, we'll make it work." We would produce to their unique requirements.

Every time we gained a new client and adapted our products to that client, we gave the specialized product its own part number. That special part number for our product was embedded into the client's inventory and ordering system. It had to keep coming back to us because our terminals were modified exactly to their specifications — no one else could match them. The margins in this kind of business are great. It's hard for our competitors to come in under us, because they can't match the customization.

Our competitors are more the Henry Ford style of manufacturers: big assembly-line operations that can't customize individual products or do small product runs. Not like we can.

However, we don't want to make products that are completely customized. For example, a made-to-measure suit is a one-of-a kind suit with all its parts customized individually. We want our customized products to be made up of standard modules that are plug-and-play, and can be put it together quickly. We chose a strategy of efficient customization. Much like Starbucks, who keep unique, limited ingredients that can be combined to make multiple permutations of drinks, we keep a minimal number of modules that can be combined in different ways to produce different results.

Ask people across the company, anywhere in Norsat, what our strategy is, and they'll say: "Customize fast."

We can build a customized product and turn it around very quickly. Our competition would normally take six months and require many extra, expensive engineering changes to achieve the same level of customization. Because we work with our modules, we have a lot of flexibility and are able to customize very efficiently and in minimal time. In our industry, where we do communications in remote, challenging arenas such as war zones and disaster areas, time is of the essence. For our customers, price isn't so much of an issue, but having the product available when they need it is.

Aside from kick-starting the revenue engine, we had to get the company on more solid financial footing very quickly. We went back and talked to our loyal customers. Because we had a great rapport with them, it was easy for me to offer them a cash discount. We said, "Pay your bill right away in cash and we'll give you a discount" — and many did. This sped up our cash flow. I went to our suppliers and worked out payment plans to make weekly payments to them. We had always maintained a solid rapport with our suppliers, so they agreed, and as a result we were able to manage our cash flow better.

There were times during the turnaround when Norsat actually ran out of cash. I had a Porsche Boxster and at one point I sold it to make payroll. On another occasion I took out a second mortgage on my home so that we could meet financial obligations to our employees and our suppliers.

It was, to say the least, a very scary time, and I worried that my car and my house were on the line. But we had one chance at success and I couldn't forget all of the people depending on me. I didn't just make one change to Norsat to make it profitable and start growing again. There were a whole bunch of things I had to change. I like to say it wasn't a silver bullet, but a dozen bronze bullets.

o o o

So, I'm sure you're wondering why I was laid off if I had so much to give to Norsat. After all, it happened just as the company was moving ahead with the development of a promising product that had the potential to reinvigorate and revive it: the portable satellite terminal proposed in my Ph.D. thesis, which I had been fully involved in since my early Norsat days as a co-op student.

Well, even though it had a lifeline with my project, Norsat soon hit another patch of financial difficulty and was looking at layoffs to save money. But I wasn't targeted just because of my salary.

Did I mention that I used to drive a Porsche Boxster? That's what I was like back then when I was young: a show-off, and even a little bit arrogant. I was not the most humble person at the time. I saw myself as one of the rising stars in the company. I had a Ph.D., had landed the big contract for making the portable satellite terminals, and had started the product's development. I was very much the star hockey player who gets the puck, drives to the net, and shoots, instead of looking to pass to a teammate who has a better chance of scoring. I rubbed people the wrong way. There was something else, too. Even though Norsat was not in great financial shape, I got the company to pay for my M.B.A. at Simon Fraser University. I was taking one day off work every two weeks for my studies. My attitude was that as one of the company's stars, of course they should invest in me and my education!

A lot of my colleagues who knew the company was in bad shape looked at me and asked, "Why aren't you working every day? Instead, you're taking time off every second week *and* the company's paying for your executive M.B.A.!" It began to look as though I didn't have the interests of the company and my colleagues at heart.

Looking back, I can understand the opinion they held of me then. It wasn't surprising that I was let go. My personal failings were detrimental to the company, to the portable satellite terminal project, and certainly to me. Because of them, having been laid off and away from the company for eighteen months, I missed the development and delivery to our customer of that first portable satellite terminal — my inspiration, my baby!

These kinds of personal failings don't happen without leading to some self-reflection and examination. As a result, I changed over time. There wasn't a single "Ah-ha!" moment; I didn't shout "Eureka!" It was, instead, a slow and steady process, with the guidance and help of others, and especially one person who deeply understood the roles and responsibilities of a CEO. That person, whom I consider my mentor, and who was on our board when I was in senior management, was John MacDonald. He told me, "You know you've become a true company executive when you start treating every company dollar like it's your own." There are many CEOs out there, and, I have to say, some that I've seen here at Norsat, who would say, "The company should pay for this, or that," or, "I've got moving expenses the company should pay for." Or they would expect limo service on a business trip. Applying John's philosophy, the question you need to ask is: Do you get limo service when you go on a personal trip? If not, then why do you need it as a CEO?

A lot of John's advice really sank in. How should I treat the company money? How would the shareholders like me to treat their money? It was part of my career growth — maturing from a simple employee motivated by self-interest and thinking about *my* benefits and *my* next raise, and can I have part of *my* tuition paid for, to asking myself, What can I do to help the *company*? In the end, these little things — such as whether I stay in a five-star hotel versus a hotel we found for a good price on an online discounter — really don't matter from a strategic point of view, but they do matter to the perception of our team and our shareholders. And that helps a company to be a cohesive unit.

Many people ask if I was scared making big decisions that affected me personally, like selling my car and remortgaging my house to make payroll. I discovered that on a personal level it was really about knowing that you have just one chance to change a company. I understood the company's financials; I did my analysis; I did my homework, and I knew I could make it work. At the same time, I figured that if I didn't take the gamble the company would certainly be gone. I didn't want to be on my deathbed saying, "What if?"

The GLOBETrekker Today

There's no such thing as an overnight success. And even when the stars align, there is still a lot of hard work, failure, and starting over.

We call the portable satellite terminal the GLOBETrekker. It has gone through many changes and generations from my original Ph.D. thesis concept to the final product. It has turned out to be one of Norsat's winners, as we have ridden the wave of growth opportunity in satellite communications.

The product has many claims to fame. During the wars in Afghanistan and Iraq, there was a very good chance that any war-zone footage that you saw on CBC, CNN, or Fox came through our equipment — through either broadcasters or the military.

We once received a call from the United States in the middle of the night. It was a customer saying it had a VIP in Afghanistan and needed our support because it had bought one of our terminals for this trip. The customer wanted someone from Norsat on standby to help with a very, very important task. They wanted to make sure we could help in case it ran into technical difficulties. And so it was

Amiee Chan with the GLOBETrekker mobile satellite terminal.

that at three o'clock in the morning, my team, who do whatever it takes to get the job done, were standing by online from our site in Richmond, B.C. Suddenly, they could see the blinking light and hear the 3-2-1 countdown to the broadcast. President Obama's face came on our screen at Norsat. It was a top-secret mission, so they couldn't tell us ahead of time who the VIP was. It was amazing! Obama was addressing the nation live from Afghanistan using our equipment.

We work with our customers. We're always very supportive. Our equipment is used for all types of important broadcasts and missions. Customers know they can count on us because the portable satellite terminal has gone through so many generations, and we've continued to make it more reliable and rugged every time.

o o o

In 2010 Norsat donated one of the first-generation GLOBETrekkers to the Canada Science and Technology Museum in Ottawa. This GLOBETrekker model was used in 2010 when thirty-three Chilean miners were trapped for sixty-nine days in the San José mine at Copiapó in the Atacama Desert. The mine is in the middle of nowhere, and our product was used by broadcasters who came to that area and needed communication. The GLOBETrekker was there to broadcast live footage from the scene as the events unfolded: How were the rescuers trying to get the miners out? Would they even get the miners out? Would the families be reunited? All was broadcast live. It was an inspiring and personal story for all of us at Norsat. Our communications products are regularly used in war situations by militaries and media broadcasters, and it was heartwarming for us to see one used for this kind of humanitarian purpose.

The Canada Science and Technology Museum has put its collection up on its website, where anyone can see images of the donated GLOBETrekker and access more information about it. To round out this story, the Canada Science and Technology Museum was the first museum in North America, and the only one in Canada, to exhibit the Fénix 1 capsule, the rescue capsule used to support the trapped miners. The capsule, which was on display for several weeks, was welcomed to the museum by Chile's ambassador to Canada at the time, His Excellency Roberto Ibarra.

What's Different About Norsat Today?

Avoiding the types of failures that Norsat experienced in the past is important to us, not only for our shareholders, of course, but for our employees and their families who are dependant on the company's success. I think we are more realistic today about the quantity and type of business that we go after. We want to balance our need to remain innovative while developing our products, trying new things, and even failing, with maintaining a cohesive company that has all its employees working toward the common goal of meeting our customers' expectations.

We start with our vision statement and our set of core values in place. We also have a ten-year vision of what the company should be. We have a three-to-five-year plan that details the capabilities that we need to be competive in our industry. For instance, we want to make sure we have faster delivery than all of our competitors. We want to make sure we have a lean culture. We want to make sure we exceed our customers' expectations.

We tie action items to the three-to-five-year plan. Then we shorten these down to a one-year horizon, and then the one-year goals are cascaded down to every employee. For example, one of our goals for this coming year is to shorten our lead time on a particular product line. It will flow down to the operations team, the production team, and the assemblers. Everyone has these goals, and everyone's bonuses are tied to the goals. This way we get good strategic alignment across the company.

I don't think there is anything revolutionary in what we are doing now, in contrast to years gone by. It may be that the magic today is simply that we are more realistic and deliberate in our decision making. We don't want anyone to be a dead weight or an obstacle to the company, and we are relying on each individual at Norsat to contribute to the company's success and to know that he or she is important to it.

2

EEPMON, A.K.A. ERIC CHAN
Digital Artisan

#FAILMORE

I met Eric last year. He is young and dynamic, and does not fit into any category. I think you will appreciate what I mean when you read his chapter. He has permitted us to gain a glimpse into what I personally believe is a real problem, how our school system tries to categorize us, how the "system" cannot understand true originality. Eric is the creator of the EEPMON brand. He is globally recognized as a digital artist on the rise, and he is contributing some of his talent to the relaunch of the Canada Science and Technology Museum in Ottawa. I am lucky to consider him a colleague with a shared vision of how amazing Canada could be, if we let it.

— Alex Benay

My name is EEPMON, a.k.a. Eric Sze-Lang Chan. I am a digital artisan and I work in the worlds where computer code and art meet. In my career as EEPMON I have worked in both fine arts and the commercial world. My exhibitions include the Library and Archives Canada, Ottawa Art Gallery, group performances at TEDx Toronto, and the *Gutai: Splendid Playground* retrospective at the Solomon R. Guggenheim Museum in New York City. I have collaborated with some of industry's best-known brands including Canada Goose, Marvel Entertainment, and Microsoft.

When Alex Benay asked me whether I would be down to write a chapter about my failures, I thought it was a bit absurd to assemble my failures into a chapter that might be seen the world over. Having given more thought to the matter, I realized the value of sharing my own personal stories and insight on this matter. While it would be much more standard procedure if I were to share my successes, my failures are far more interesting, and I really do think you will be much more entertained! Writing this was a great exercise, as it was a chance to reflect on my more than ten-year career as EEPMON — my artist alias and brand.

When we focus only on people in the limelight, the optics are that they are amazing, successful icons of Canadian society. Naturally, curiosity sets in and we wonder in the classical manner of the five *W*s and one *H:* Who did they know to get there? What did they do? Where did they go? When, Why, and How did they

EEPMON's open data weather performance painting, "INTERSECTIONS" at hpgrp Gallery NEW YORK.

get there? I can think of no better way to provide this insight than with the story of my failures!

Failures are easily remembered because they hurt. I consider my failures to be battle scars that still burn to this day. At the same time, these scars are key points in my personal and professional development and have brought me to where my brand EEPMON is today.

I've divided this chapter into ten sections. Let's get started!

Failure from the Start

I had a hard time focusing and learning during my elementary school days. I was slower than everyone else. I had to add another year to my schooling, after which I felt distanced from the others because I was a year older than my classmates. I took special language classes because I had trouble pronouncing words correctly. I was behind in class, had a very short attention span, and became bored easily. I could not grasp the concepts the teacher was trying to pass on. My marks suffered horribly and I was frequently in danger of receiving that dreadful *F* grade.

High school wasn't any better. I went to Canterbury High School in Ottawa, known for its wonderful arts program. However, I still didn't really fit in with my peers. When everyone was painting with acrylics and pastels, I was painting in Photoshop. When my classmates were making a series of ceramic bowls for a pottery class assignment, I created a series of Japanese anime-inspired mechas. I seemed to go in the opposite direction to everyone else.

I was immediately drawn to computers when I recognized that they could be used to create art. As an avid video gamer, using Photoshop to create pixel graphics was like weaving magic. I was obsessed with it and I would try to spend as much lab time as possible on the computer (we only had one!).

I miraculously survived high school despite my disappointing grades. I emerged angry and confused. I thought to myself, "How will the eyes of society see me now?" I loved making art but was told that a career in it would be suicide. I felt the people telling me what I should be doing with my life were pulling me in all directions. On top of that, there was the pressure for me to satisfy society in the way that I understood it: I thought the path to success was to get a university degree, get a stable job, buy a home, get married, and live happily ever after. At that time and with my low grades, none of these even seemed vaguely possible. It even felt like my art was falling in a downward spiral.

With high school done, I considered my next steps. I had not been accepted into the university art programs to which I had applied, so I was left with two choices: Eric Chan pursuing Computing Sciences at Algonquin College or Eric Chan in the University of Ottawa's Economics Program. As I still loved computers, I chose Algonquin College. I look back at that as my first successful life decision.

Failure to Be Honest to Myself

When we are young we have dreams. We have excitement. We have passion! But then something called "adulthood" comes in and we are told: "Grow up!" "Get a good education, get a job, get married, get a house, and live happily ever after!" were the words I heard. So who could blame me for perceiving that as the only version of success? There was no one telling me otherwise. There was no one telling me that I could do whatever I wanted. I was surrounded by complacency in acceptance of the status quo. So much complacency that it was hard to even breathe.

I digress. But I think there are those among you who can relate to this. In our late-teenage years we are told to take responsibility and figure out what it is we need to do in life that can guarantee stability. So what do we do? We completely shut away the things that give us joy: our hobbies, our passions, and our dreams, as they tend not to be things that we see as relating to safe careers. "Oh. Drawing. That's nice. Can you make a living from it? What's your *real* job?" people ask. We create façades and search to find definition in our lives to allow us to fit within the accepted social fabric.

I surprised myself by quite enjoying my time at Algonquin College. People were more mature and more willing to accept difference. I started to fit in for the first time in my life. In my second year I had the opportunity to secure a co-op placement at a software company known for its graphics programs. It was my first step into the working world — as a Quality Assurance Specialist. I got paid and was able to do the job well. My confidence grew and real work was not as scary as I thought. There was a particularly close-knit group that I walked by every day — the User Interface / Experience Design team. They were the ones who created the graphical user interfaces for software, and I thought their work was really cool. I set up a meeting with the team's manager on my last day to see if there would be an opportunity to work for them on my next co-op placement. I showed her some of my graphic art. She stared at it and I vividly remember her saying to me, incredulously, "*You* did this?"

In actuality I didn't think much of the pieces because they were something I did on the side, and art was already something I had accepted would be a hobby, and not a for-real career. So when I said "Yes," I was ready for the bad news. Instead, I received great news. She was so impressed she immediately signed me up for my next co-op position as a User Interface / Experience Designer!

The "Eureka!" moment hit me. "Eric," I said to myelf, "how dare you try to suppress your own creations! It has been your ambition all this time! You always felt that your work wasn't recognized or appreciated by your peers, and yet you impressed a professional so much that you got the job that you wanted! What is going on? How could this be?"

It was then that I finally saw the signals the universe had been sending me and realized the direction I needed to go in. Another realization dawned on me: I failed myself! My failure was in listening to too many people. How dare they tell me what was and what was not! I had forgotten to listen to myself!

Looking back now, I see I needed to be pushed to extremes before I could truly break out from the mould. That particular experience has been seared into my memory — it is one of my battle scars. After Algonquin I went on to further my education at Carleton University's Interactive Multimedia and Design program, during which EEPMON was born.

o o o

To this day, while I am accepted by many, I continue to be the odd one out. I do not fit in any particular group, and that is totally fine. I have *embraced* my difference!

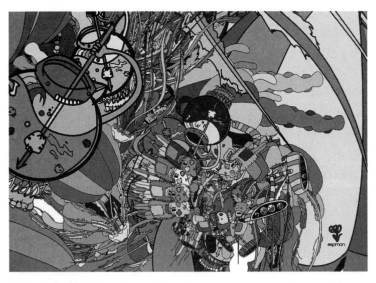

"The Delightful Tea Party Bots," 36 x 24 x 1.5 inches, giclée on canvas, 15 editions.

Failure to Recognize My Artistry — Time to Own It

When I introduce my work to people, many ask if it was hand drawn. [*Insert face-palm motion now.*] Of course it is! Drawn with a Wacom tablet in Adobe Photoshop, Illustrator, and/or generated by code that just so happens to be typed by my fingers, which are part of my hand. Their reaction is priceless! It is a look of bewilderment in combination with slow nods. It is disappointing that those outdated modes of thought on what constitutes a legitimate work in the art world still continue to proliferate.

We shouldn't get trapped into the same old dogmatic thoughts about what constitutes "legitimate" art. "Is it hand drawn? Is it painted? How long did it take to make it? Was it done on a computer? Was it done on a tablet device or made with technology? That is blasphemy! Forget it! That's not art!"

But really, now? By definition, a pencil or a paintbrush is also a technological tool. You cannot judge a piece of art by what was used to make it, or how it was made.

Now that I think about it, that is probably why I started EEPMON. I needed something that represented *me*, an artist of the digital age. Something that speaks truthfully about me and that no one can tell me to change. Even if they think they know better, they can see what EEPMON has accomplished and they will back down. EEPMON is my own pirate ship setting sail into the world — on my own terms. It is time that I own it!

EEPMON is my thing that sparks joy and adventure. EEPMON cannot be quantified. EEPMON cannot be categorized. EEPMON is holistic. EEPMON is my DNA.

And not just me. I have met many others who, despite the pressures of society, continue to pursue the very things that excite them — so much that they will live and die for them. We share similar struggles, stories of failures and the hustle for success. We persevere and empower one another. We are unconventional.

Traditional modes of education didn't work for many of these others, and they didn't work for me. I am a very hands-on individual. I need to experience in order to truly understand why things work in certain ways and why they do not. I am not by-the-book. Letting others tell me what to do disrupts my focus.

So, a declaration! I AM AN ARTIST OF THE DIGITAL AGE. A DIGITAL ARTISAN. TIME TO OWN IT.

Case Study: Failure in Tokyo

Jump to 2009, Tokyo. Having successfully exhibited my open-data performance paintings "INTERSECTIONS," the curator of my show introduced me to a well-known Japanese musician. He loved artistic collaborations involving improvisation. Both the musician and I felt that this would be a great opportunity to explore music and digital art in a performance. We agreed to meet the following year, and even secured a venue.

Our venue was at the newly built Yamaha Ginza — the flagship store for Yamaha in Tokyo. Of course I was very excited. This was the moment that I had been dreaming of: EEPMON was finally making international strides, travelling halfway around the world to work on artistic and creative projects. How cool was that! I worked so hard to achieve the dream. This was the moment I had been waiting for. In just five years, EEPMON was going to make things happen in Japan.

For the performance I developed several programs. The first I created generated shapes and colours based on the high and low pitches of the sound received by the microphone. Using a Bluetooth keyboard, I assigned a different key to a different set of generated visuals. It was my very own instrument. The second program was inspired by one of openFrameworks founder Zach Lieberman's "Reflection Study," involving a lightbox with a webcam facing down on it. By placing cut-out cardboard shapes onto the lightbox, the webcam would scan the surface and project the image in real time. The cool part was that visual effects and sounds would be generated by the program based on the shape of the object. The more objects placed onto the surface, the more elaborate the installation/performance became. I thought I could do something similar but with my own spin.

Everything leading up to the performance seemed to go smoothly. What could go wrong? But things started to change. The musician decided to bring another person into the performance. I did not know who this person was and I wanted to stand my ground and tell him "No," as I should have done. I have never met such an arrogant, egotistical, demanding person as the new musician. He was a snob, criticizing everything and thinking he was the big shot. The curator and I were at a loss. I was disappointed and upset, thinking that this was not what I wanted out of a collaboration. This was not what I expected as my dream. I had travelled halfway around the world for this? Unfortunately, I couldn't do anything about it. The options were the show with them or no show at all.

The day of the performance came, and so did the multitude of failures that soon followed.

I forgot to turn off my screensaver. As soon as the musicians started to play, the Mac OSX space screensaver turned on. Improvisational music with space-themed screensaver — Good job, Eric. Even worse, I had trouble turning it off because I was up on stage with my Bluetooth keyboard. Panic ensued with all audience eyes on me. I had no choice but to run back to my computer and manually turn the screensaver off. There went the first ten minutes of the first performance. *Utter failure.* The sting lingers to this day. I am cringing about it even as I type this!

I failed to double-check my equipment. I failed by assuming that the technical setup would be easy. I failed to turn off that damn OSX space screensaver ... I failed to stand up for myself and voice my opinions. I failed the audience who came and paid to see our show. I failed my dream. Finally, I failed the curator. It ended awfully.

After the ordeal I was in shock. I will never ever forget it. I still think about it and it continues to sting! It was my dream gone bust. I thought I had it. Important people whom I needed to impress were watching that day. It was my moment to shine and show them what I was capable of. They were brought together through the kindness of my curator and my curator's connections. I worked hard to bring them all together — and then ... *I fucked up.*

I wanted to hide and not show my face to anyone. I wanted to go back home, lock myself up in a room and reflect. What an awful experience. I know that people have forgotten about it and I have only myself to blame for continually bringing these memories back to the surface. But maybe this is a good thing. A good thing, because it is, in fact, an experience worth remembering. A hard-learned experience that humbles me, keeps me grounded, and builds strength. Never assume things will all go as smoothly as you hope. So many variables are at play, and you have to double and triple down on every detail, especially if it involves people paying to see you. (Ugh —I just felt the sting again!)

Case Study: Failed Anticipation

I have learned how to be Zen at almost everything. I have learned to control my temptations and indulgences. I have learned to fail in order to achieve. In 2010 I received an email from some people at a New York City agency. They wanted me to collaborate with a fashion brand. I can remember how excited I was, and how I thought it was my moment: art, collaboration, New York City, and fashion! Days went by, and then weeks, with no communication. I was becoming worried. I wrote back to them and they said they were still working out the project logistics. My spirits rose again. More months went by. Six months later they told me that they had used another artist for the collaboration.

It felt as though a knife had stabbed through my chest. I went from high in the clouds to down into hell. How could they have stood me up that way? They approached me! They wanted me! I was sure it was for certain!

I learned an excellent lesson: never, ever assume that a project is a go unless documents have been signed and the product is actually released. Until then, it is all talk. Always remember that you may be one of many candidates. I now say: "I'll believe it when I see it — and not until!"

This did not happen to me just this once. It has happened more than once with different agencies. So I have also learned to take a Zen approach to opportunity, because opportunities are only opportunities. They may or may not lead to anything. Just keep calm, be thankful and appreciative, and hope that something will happen. No need to get excited. The offer is a positive sign, but that's about all. Exerting energy on a matter that is not guaranteed is not worth the exertion. Just be. Through this failure I have learned to appreciate what I have in the here and now. The future is all potential. Nothing is certain. Respect the unknown.

Observation: Why Are We Afraid of Failure?

I may be generalizing here, but I often wonder why we humans dislike failing. As I see it, our society has been structured in such a way that when we attempt to do something that is out of the norm, it is immediately open to scrutiny. A "You do not fit in with us" group mentality sets in, and the group suppresses those who desire to try something new and different. That's because, for most, difference is dangerous and impacts their ability to be accepted.

Let's take one step back with this question: Where did this notion of failure come from? How did it become part of our human psyche? We spend at least half of our lives working at things called careers. We are either commended for our good intentions or rewarded appropriately with money or recognition. Society is very much structured around reward systems. We are no different than the mouse in an experimental lab conditioned to do tasks to get rewards. If we do not do what it is expected, then there's no reward. We feel miserable because we notice our peers being rewarded for doing the "right" things. Pressure builds, and we conform and do what society has conditioned us to do: follow the rest, choose the outcome with the least resistance, and be rewarded. This is complacency. The world's economy is built on everyone working and trying to climb higher to grab the bigger goodies at the top.

I think people associate failing too closely with negative outcomes, and this has infused our culture. Sometimes I wonder if fear of failure is what makes the world go around. Seriously! We worry that if fear of failure were not ingrained in our society as a deterrent, our entire civilization would collapse. Our economies would collapse. It would be total, utter chaos. But I think fear of failure is there to keep us complacent and obedient. Those who have ambitious dreams to realize, and the conviction to pursue them, must stand up against this fear of failure. In fact, I encourage you to fail more.

Observation: Canada and the Arts

I'll say it now: I think that art institutions also tend to stick to the path of least resistance. Have you noticed? It's the same people picking the same pieces to be shown in the same places. They speak only to a small fraction of our Canadian population. Our nation has so much more to offer!

And so, people (like me!) — not of that "small fraction" — have to do things their own way. I have a background in computer science and interactive multimedia and design, and yet I call myself an artist. I've always viewed myself as an outsider and a disruptor. That's why for the past ten years I have been doing things my own way and building EEPMON into what it is today.

People will ask me, "Why don't you apply for this grant or that residency? Look at this person and that person! They got in! Why don't you?" Truthfully, I don't give a shit. I know it would be a waste of time. I see it this way: I can divert the energy I would put into an application with very low prospects and instead direct it toward getting a project all on my own with very high return. There is value doing it this way, too. I've been working like this for pretty much all of EEPMON's career — the grassroots way. It has always been the grassroots way.

Perhaps this is my entrepreneurial side speaking. I hustled hard to get my opportunities and I've been able to bypass the middleman. The digital age allows this. The digital age is a leapfrog agent. This age will not wait for laggards. Industries (including the arts) that wait too long to adapt to today's reality will suffer. Suddenly it will be too late and too costly for structural reforms, and they will crash. We have seen this happen many times already. Take a look at Blockbuster Video, HMV, and BlackBerry. Blockbuster waited too long to get into the streaming business; it even had the opportunity in the early 2000s to acquire Netflix. HMV suffered huge losses as consumers found it much more convenient to purchase and stream music online. BlackBerry got comfortable and waited too long to bring touch technology to its devices. Even though I am referencing media companies, the paradigm holds true for the arts. If they do not change, they are doomed. By playing it safe we are stalling. Rise up! A new era of creators is here, and it is now that we must acknowledge them!

That being said, it does appear that things are changing. Now, after Canada's 150th, I can see progress being made. The Canada Council for the Arts reformed its funding model to be more inclusive of other forms of media that at one time may not have fit within its rigid categories, and it is now looking into future. Computer graphics are finally starting to be an accepted medium in the fine art world.

Overall, Canada is showing renewed optimism. We are more open, more outspoken, and more aware. By the time you are reading my chapter, I think you will agree with me, too.

Canada's traditional art scene must realize that the definition of "exhibition" has changed. Exhibitions are not only housed in physical galleries. Whether it is on a wall, on a fashion garment, on the Web or even on a coffee cup, an exhibition is an exhibition because it is doing what it is supposed to: exhibit, and bring the art to the people. *All people.*

I'm the new breed of artist — one who is able to make art, technology, and business intersect. You should get to like me, because there will be more of my kind. Those kids playing Pokémon and listening to video-game music for fun? Keep an eye on them. They will become the next generation of innovative creators. I hope for the sake of the art establishment it will be prepared.

As a digital artisan, my potential is limitless. I have lofty goals. Global goals. My sights are set on the big fishes in the oceans — not in the ponds. I work with the doers, not with the talkers. What were once my dreams have now become my realities. Because, and I'll say it again, the digital age encourages this. I'm ready. I wonder if Canada's art scene is ready, too.

#FAILMORE

The *F* grade may be seen as utter failure, as unacceptable to the commons, rendering one an outcast from the group. The *F* grade has been ingrained into our subconscious minds with negative connotations. If we are bearers of this dreaded letter, we try not to show it, nor talk about it; we are ashamed of it.

The fact of the matter is that we are all just human beings. There is the pursuit of perfection, but there is no such thing as perfection. There is no such thing as the ideal human. We are beings in constant flux, moving through this enormous universe, and we forget how insignificant our actions are. The person who is making a big deal out of your actions is *you*. Have you heard the saying, "You are your own worst enemy"? Even if you feel really bad that you failed, *it is no big deal*. The world and the universe keep moving.

Instead of dreading failure, see it as a challenge that life throws at you. Failure guides us to better ourselves. Challenges give us measures to see how far we can push ourselves. Sort of like setting the limit breaks. When you have successfully broken that limit, you have levelled up and are ready to take on the next limit. You have become stronger, and you now have the know-how to better deal with the next challenge. Have you ever stood back and wondered why, of all the things you have done, the failures seem to be the ones you remember most? That is because through these lessons you have become stronger! From failure you will find strength.

So you say you can't handle failure? As a good Canadian I will have to say, "Sorry!" How boring would life be without failure? What will you do? Sit there not trying? A big #YOLO to you. Go out there, do something risky. We Canadians seem to be just fine being complacent and we seem to be satisfied with the status quo. But we as a nation have the right ingredients to take charge and make a significant impact in this world. So we need to fail more. Failing more creates new opportunities and new experiences. Failure is at the heart of innovation.

Besides, what is failure, really? Failure is just a term that society applies if something does not seem right, or goes against conventions.

So we must fail more. When we do more we fail more. Failing will allow us to build deeper understanding with ourselves, our communities, our industries, and society. It needs to be an acceptable point of conversation, disseminated and calibrated. Failure is just the beginning; we use it to finesse and tweak so that our next iteration will be better. I sum it up pretty much like this: our unexamined dogmas keep us safe, keep us complacent, keep us ignorant — but with the right amount of entertainment to keep us distracted from discovering our full potential and ultimately our truths.

Don't be afraid to get out there and do your thing. Failure is there to test us. After one failure there will be another. We have to acknowledge failure and embrace it to grow. Failing is what builds character, strength, and confidence. One of my mantras is, "If it were easy, *everyone* would be doing it."

Revelation: What the Forest in Kobe, Japan, and the Unknown Taught Me

In 2012 I was in Japan, exploring city-to-city with my very convenient Japan Rail Pass, which allowed me one week's unlimited access on the Shinkansen (bullet train). I decided to visit Kobe — a beautiful Japanese city and one of the first seaports to trade with the West in the late 1800s. And of course it is known for its renowned Kobe beef! In a typical EEPMON move, I managed to get myself lost as I wandered around the city, eventually ending up at the foothills of Kitano. Even though the sun was setting, I decided to trek up the mountain.

Walking up the mountain through the forest I can hear the wind slowly ruffle the leaves. Looking up, all I can see is the silhouette of the leaves imposed against a dark blue sky.

I walked in feeling like I was entering into an ephemeral and transient space. I was alone walking up this path ... behind me it was pretty much black. The darkened blue sky was my only guide. Eventually I ended up at a lookout with a view of the city. It was beautiful. I decided that I would just stand there and take it all in, and, as any artist should, undertake some reflection. And then it happened. Out of nowhere, over the horizon, the supermoon appeared. It was gigantic and strikingly red. It was figuratively and literally out of this world. I felt that the moon was speaking to me: "Eric, you have made it this far because of your perseverance and your conviction. Despite the obstacles that life has thrown at you, your ability to fall and rise again speaks to your courage and commitment to

EEPMON sporting the EEP snapback cap and the EEPMON leather jacket.

make your dreams come true. You are here because you have struggled, you have failed, and you have endured. I, the moon, acknowledge you. Keep going, Eric. You are here for a reason. You have found me. Now you can continue moving forward."

What were the chances of my being face to face with the supermoon at the lookout view in Kobe? The experience was a spiritual gift. The moon reached out to me, speaking to my subconscious.

The revelation: failure and success are like yin and yang. One cannot exist without the other. You cannot grow if you only see one side of the story. You must experience it in full. Not giving up and continuing on your own path is a sign of strength and courage: if I did not fail in my Tokyo performance; if I did not fail in getting that lucrative fashion collaboration; if I did not succeed in winning a design award; if I *did* succeed in creating new artworks, *did* succeed in my collaborations — in all of these I demonstrate my conviction. That splendid view with the supermoon in Kobe confirmed it. A moment of self-actualization.

#FUTURE

More than ten years have passed since EEPMON's first steps into this world. Looking back through my successes and failures, I can connect the dots leading up to where I am today. It is fascinating to see how they are all intertwined. *Everything* that you do in life has a reason, and there is a reason for everything.

Where do I see myself going? It used to be that I would draw for the sake of drawing — and I still enjoy it, but I also now see that I have a much greater role to play on the world stage. I have reached a milestone and I am proud of what I have accomplished, but I know the journey is long and I am far from finished. I like to think that I am just getting started! I see new horizons to be reached, new goals to be met, and I will take my abilities as a digital artisan to a whole new level. So, what are you waiting for? Let's get up and get it!

On that final note, please do follow me on my continuing adventures via my handle EEPMON on Instagram, Twitter, Facebook, and WeChat!

3

TOM JENKINS

Chair of the Board, OpenText Corporation

When One Door Closes, Another Opens

I've known Tom Jenkins for many years. I first met him during my first stint with the Government of Canada, when he was Chief Strategy Officer and Executive Chair of OpenText. Tom has since "retired" and now serves only as Chair of the Board of OpenText, Chair of the National Research Council, Chancellor of the University of Waterloo, etc. — a typical retirement for most of us, no doubt!

To me, Tom represents the very best of Canadian leadership. He is a proud Canadian, and he is involved in public policy, charity work, and national innovation and productivity discussions. I learned a tremendous amount about leadership, strategy, and management from this man during my time at OpenText. Even to this day I consider him a friend, mentor, and colleague, and I am extremely excited that Tom has agreed to share some of his thoughts on the topic of failure.

— Alex Benay

When one door closes, another opens; but we often look so long and so regretfully upon the closed door that we do not see the one which has opened for us.

— Alexander Graham Bell

The first half of this chapter provides a quick look at the early stages of the internet and OpenText Corporation, how and why it grew, the opportunities it seized upon, and the huge bump in the road that surprised and personally affected me. The second half will deal candidly with my thoughts on what this failure meant to me and on the bigger question of what failure means to Canadians and to Canada as a society.

o o o

Today, we take the internet for granted. With a few keyboard strokes or taps on our mobile screens, an internet search engine takes us anywhere we want to go on the World Wide Web. This is a story of the early days of the search engine and how I was part of a very public failure on the World Wide Web during the dot-com boom of the late 1990s. I was not alone. There was a lot of failure going on at that time precisely because there was intense innovation happening. It was a remarkable time that I have never experienced again.

The internet is a public network. However, the same technology searching the World Wide Web also searches intranets — those private, behind-the-firewall electronic networks of information and databases of governments, corporations, institutions, or any organization that chooses to have one.

A humorous story involving an early intranet — and a very small intranet at that — involved the desktop computers at the University of Cambridge in England that were networked to what is claimed to be the world's first live webcam. This webcam watched a coffee pot so that coffee drinkers whose offices or desks were more than a few steps away from the hotplate could check first to know whether it was worthwhile to walk over to the pot. This webcam network is an example of an extremely small intranet or private network behind a firewall not open to the public. Today, intranets account for 98 percent of the electronic information in the world. Is it any wonder, then, that intranets are collectively known as the "deep Web"? Understanding that is a big part of my story.

o o o

Today Canada is one of the leading countries in the management of intranets, thanks in part to OpenText. Back in the earliest days of the World Wide Web in the 1990s, Canada had a chance to create the first dominant internet search engine. In fact, for a few years between 1995 and 1997, a Canadian internet search engine called the Open Text Index was one of the most widely used in the world. However, it all ended in failure and disappointment. This failure would have a lasting impact on me personally, and I was caught by surprise by the magnitude of the reaction to the failure of the Open Text Index, and by my own personal emotions about it.

What went wrong?

Let's start at the beginning. In 1991, before the internet was widely popular and before the World Wide Web had been adopted, three researchers at the University of Waterloo, in Ontario, devised a clever method for searching the *Oxford English Dictionary* (OED). The OED at that time was the definitive text on the English language. The twenty volumes took up an entire shelf of a library and the memory required to store the OED was an astounding six hundred megabytes! At the time, a five-inch floppy drive (like an old record platter) stored less than one megabyte, so six hundred megabytes was significant! The thinking was that if OpenText was able to index an entire dictionary, which was really a database of words, then how much harder could it be to index the Web, which at the time was about one gigabyte, just a little bit bigger? However, this was considered an ambitious goal since OpenText had fewer than twenty employees. This OED research initiative eventually led to the development of an index of the World Wide Web called the Open Text Index, or OTI for short.

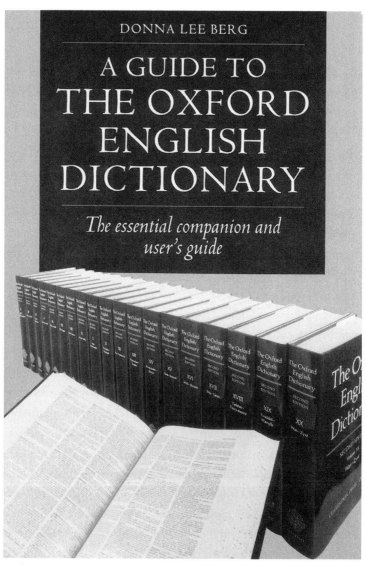

The *Oxford English Dictionary.*

At the time, OpenText's indexing technology was leading-edge. One of its claims to fame was that, owing to how the technology was constructed, as the size of the database grew the indexing function was able to scale to match it. This was very important in a world of very limited memory. Developed in conjunction with an internet service provider (ISP), the OTI used the ISP's server for network access and high-bandwidth facilities to perform the indexing. Content was gathered using a new tool called a "crawler," which went around the internet to every Web address and every link from each website. At the time, there were only twenty-five thousand Web pages. Today, there are more than four billion pages in the public Web. Most estimates are that the public Web is less than 4 percent of the entire Web, although no one knows for sure, this suggests that there may be as many as one hundred billion Web pages in the world today! It's important to understand the shift from public Web to private intranets as searching the Web developed.

Back in 1995 OpenText announced its OTI service designed for indexing and searching the World Wide Web. Free to anyone on the internet with a Web viewer, the Open Text Index was a full-text search tool that included an internet agent capable of continually visiting, reading, and indexing every word in every document on the internet.

The internet was creating new opportunities, and OpenText management moved quickly to snap up expertise and expand its market opportunity, in both the public sphere and private internets that was were being formed inside companies and governments. During these early days, OpenText found synergies in staff and technologies in two nearby companies. The first company, called Internet Anywhere, was located in Waterloo, across the street from OpenText. These engineers understood the use of hypertext markup language (HTML), which was the typesetting code of the new platform known as the World Wide Web (WWW). The second area of expertise required was the language of transportation of the bits and bytes over the internet. This language is known as hypertext transfer protocol (HTTP), and this knowledge was found in nearby Toronto at NIRV Centre, one of the first internet service providers in the world. NIRV even owned the Web address internet.net. The computer scientists at NIRV Centre knew how to "hop" the internet very efficiently, which saved time and money for things like crawlers.

The Open Text Index was really created by combining three different engineering teams: the University of Waterloo team, with the original search algorithm for the original *Oxford English Dictionary,* the Internet Anywhere engineering team, and the NIRV Centre communications team. Together, they built the most successful internet

The first OpenText website.

search engine up to that point in the history of the World Wide Web. For a few years, the Open Text Index became one of the most used websites in the world. All of this was designed in Canada in the Waterloo-Toronto corridor. However, the software development and talent acquisitions all required investments. OpenText needed a source of capital that extended beyond venture capital in order to finance the rapid pace of innovation. The company had to consider a public offering.

OpenText Teams Up with Yahoo!

In 1995 California-based Yahoo! was a directory that listed websites on the internet. Yahoo! founders Jerry Yang and David Filo were *manually* creating the table of contents of the World Wide Web, similar to the table of contents at the front of a book. However, as the World Wide Web grew, it soon became impossible to keep up and gaps were appearing. When people went to the Yahoo! site and performed a search, about 20 percent of the time they came back with no matches.

What they needed instead was an index from the "back of the book" that took every word of the entire World Wide Web and listed it alphabetically with a pointer back to its original page. OpenText gave them one. Instead of just listing sites, OpenText's search technology allowed Yahoo! to search every word on every Web page — over a billion words, over fifteen million links. So a deal was struck. Instead of getting zero results, the Yahoo! search would automatically be handed over to the Open Text Index so that a result could be found.

The relationship proved very profitable. OpenText became a large shareholder in Yahoo!, an investment that would substantially help OpenText's balance sheet and help finance its growth into new areas such as the private internet (also called the deep Web). In an article in *Report on Business*, September 19, 1995, OpenText hinted at things to come: the company "expects its real profit in the relationship [with Yahoo!] will come when corporations see how powerful its search technology is and want to customize it to search their own data pools."

OpenText, through its Open Text Index, was doing more than one million queries per day, which at the time was an unheard-of number.

Tom Jenkins (right) and Jerry Yang (left) launching the Open Text Index in New York City in September 1995.

Everything ran smoothly for a couple of months, with the load from the Yahoo! service increasing steadily. In fact, Web content was increasing daily, doubling every six months and constantly requiring expansion, software, and hardware upgrades, just to keep up.

Then it happened. On December 12, 1995, a huge storm blew in from the Pacific, packing hurricane-force winds of 160 kilometres per hour and torrential rains. The storm swept over the entire San Francisco Bay Area, knocking down telephone poles in its path and leaving more than one million Bay-area residents without power. Worse still, due to the widespread damage, it looked as if some places would be without power for days, possibly weeks. The timing of the storm and possible outages were particularly bad because OpenText had been preparing to go public, and having the Open Text Index down for a few days, let alone for a few hours, was unthinkable.

What was OpenText to do?

OpenText moved into action quickly. The local staff in California and a new IT manager (in his first week on the job) rented a four-thousand-watt gas-powered generator, and within a few hours they had a single server back on the air.

In late 1995 OpenText had an insight into the future of the internet. We had begun an aggressive strategy to build search engines for both the internet and intranets. It was increasingly clear that simply finding things on the Web was not enough. We reasoned that users would want to do something with the files that they found by repurposing the documents and then distributing them once again. At the time, this repurposing software was known as integrated document management (IDM). This reasoning was the impetus for OpenText to consider merging with an American company that made this kind of IDM software — Odesta Systems.

The merger was complicated due to relationships that Odesta already had with Microsoft and Apple — two emerging giants. There was a huge risk that revenue from Odesta's product (called Livelink) could falter if other corporations wouldn't adopt the internet search engine for their internal documents. It was a leap of faith that OpenText and Odesta took together.

In the world before Facebook, Google, Box, Instagram, and Snapchat, there was Livelink. Livelink allowed people to work together over the internet for the first time. This had many advantages. In 1996, Livelink was one of the first products that enabled a new business process called knowledge management. Livelink enabled people to work together, sharing and using files all at the same time over the internet. Previously, that had only happened in local area networks (LANs) from companies such as Wang and Novell. Livelink changed all that. It allowed businesses to work together using the internet.

Odesta was the most important acquisition OpenText had made — in reality, more of a merger of companies of equal size than an acquisition. Given their geographical distance (Odesta was located in Chicago), integrating operations was a challenge. With perseverance from the staff of both companies, the merger was successful, and indeed established a model for all future integrations. Odesta was responsible for rounding out the core functions of the main product that would be sold by OpenText to corporations for their intranets. It was an essential part of the growth of OpenText.

The period of 1996 through 1997 was perhaps the most dynamic period in the history of the company. The market for the internet was in full swing.

At this time, OpenText underwent a tremendous amount of change. The company had two active products: Latitude, for searching within corporate intranets, and the Open Text Index, for searching over the public internet. Latitude was sold as a software product, while the Open Text Index was available for free and its revenues were generated from advertising.

In January 1996 OpenText completed an initial public offering and became a publicly reporting corporation. It went public on the NASDAQ with an offering of 4.6 million shares priced at $15 each, netting $61 million. The company expanded to more than two hundred people and had annual revenues of $10 million by the end of 1996. Five offices were located throughout the world, with the principal locations being Waterloo, Chicago, and St. Gallen, Switzerland. Corporate users were numbering in the hundreds, but the brand of OpenText was now reaching millions of consumers through the search engine.

The Competition for Search Engines on the Internet Heats Up

The search engine market soon became hyper-competitive, with too much investment flowing into too few opportunities. It inevitably resulted in a shakeout. Not all of the early search engines could survive the intense requirements of people and computing resources to provide sub-second responses to search queries that consumers demanded. All the original search engine companies (OpenText, Lycos, InfoSeek, and Excite) on the original Netscape home page of 1996 would eventually fail at some point in the ensuing years. Major multinationals such as Disney with its GO network, and Digital Equipment Corporation with AltaVista, invested hundreds of millions of dollars into the area and overwhelmed all of the startups. In the end, they did not prevail, and a new startup called Google would prove to be the winner, due to its tactics, superior investments, and technology.

Our mistake at OpenText was to underestimate the rate at which the internet and the World Wide Web would be adopted. The Web simply proved to be too popular too fast! The only effective strategy in that situation was to be a sole-purpose company and pour all of the resources into the fast-paced sprint that was happening.

From the onset of OpenText entering the internet search engine market, we had prepared two alternatives — one that involved the public consumer search engine and one that involved deploying a

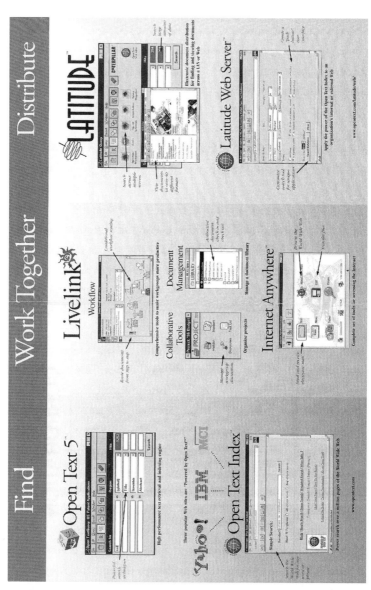

The OpenText IPO explanation of internet and intranet.

search engine inside a private intranet behind a firewall. The two approaches required different implementations of the same technology. A public search engine is a service that can be taken down overnight and replaced with a new version. A private search engine for an organization's private network would require software updates and integration from half a world away. These two markets required very different implementations. Our competitors had, by-and-large, chosen to create only the public search engine and had not developed the implementations needed to deploy this engine at thousands of other servers — a complex undertaking requiring a lot of customization. Ultimately, OpenText had to decide whether to remain as a public search engine or opt for the less competitive private search engine. We explained this to our investors in our initial public offering (IPO). Our natural Canadian caution to leave options in case our first strategy failed to work saved us.

As the demands of resources and people accelerated with the internet, it was no longer wise for OpenText to fund both implementations, and thus the company was faced with a choice of which market to pursue. The board of directors chose the safer, longer-term option of private intranets. The switch was rather seamless. Company strategy was solidified at the end of 1996, as the software model for deploying intranet applications inside firewalls became the sole focus of the company. OpenText exited from the consumer part of the internet by selling off its technology and Web traffic to the Mining Company, which would later be renamed About.com (today a division of the *New York Times*).

OpenText did a great job of completing the pivot and moving into the private internets, with all of its resources focused on solving problems inside the firewall. This led to a period of rapid innovation, in which we added even more features to the original search engine.

However, that did not prepare me for what happened in the stock market. While OpenText had always informed investors that it had planned to pursue both the public search engine and the private search engine markets, the investment community more readily understood the public search engine, since they could see it and use it.

When we announced that we were no longer going to be the search engine for Yahoo!, the stock market reaction was immediate and severe. This then affected all of the investors in the stock in a domino effect — and the share price collapsed from $20 to $2 in less than six months.

OPEN TEXT™
CORPORATION

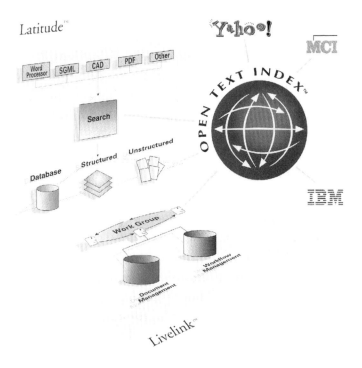

The OpenText IPO explanation of the intranet. "OpenText's integrated products enable an organization's users to find information, work on it together, and distribute it locally or over the World Wide Web."

The result was layoffs and closing of our offices in California and Paris. I was the company CEO at the time and saw this as a very public and personal failure. To put the $2 share price in perspective, the company had $4 per share in cash at the time! Someone could have bought the entire company for $2 per share and received $4 per share in cash! Today, the same share would be worth more than $300, or a cumulative, compounded return of more than 30 percent per year for the past twenty years. The market capitalization of OpenText is double the market capitalization of Yahoo!'s recent sale. The road for OpenText proved to be a complex one in those years, as the company learned from its failure as a search engine company and made the pivot to the deep Web. Another door had opened.

o o o

The second half of this chapter talks about my experiences with failure and about the lessons that I learned.

Let me tell you what happened, because it was a near-death experience. Moreover, it was widely publicized at the time. As a startup company, we were the first internet listing on the Toronto Stock Exchange, and only the third internet listing (Netscape was first, followed by Open Market and then OpenText) on the NASDAQ, the New York exchange specializing in high-tech offerings. OpenText was a bellwether for the dot-com era. We, along with all the other startup companies, went after the same thing — the search engine of the internet.

We saw an opportunity. At the time, Google did not yet exist. The original five companies were all startups from universities. In addition to the Open Text Index, there was Excite from Stanford University, Lycos from Carnegie Mellon University, Infoseek from the University of Massachusetts, and AltaVista from Digital Equipment. The other members of the original five are all gone now. Only OpenText survived.

It did so because it recognized its failure and acted before the company and the technology were destroyed by bankruptcy.

OpenText originally set out to be the primary search engine of the internet and it failed. From a Canadian point of view that's very important, because part of the dilemma we have in Canadian innovation, invention, and competition is that we as Canadians shy away from failure. If somebody fails, we avoid looking that person in the eye. The best word you can use is *stigma*. There's a stigma in Canada around failure. And that's why this narrative is such an important one. The stock price collapse was a very public, and a very personal, failure. At the time, articles about the collapse of the business model and the abandonment of the search engine were appearing in various newspapers — not unlike the articles recently published about BlackBerry, Nortel, and Valeant. Corporate failures are very public and make the news. My friends in Waterloo, my family, and other people that I know in Canada lost large amounts of money on the stock that was OpenText. When you take the garbage out to the bottom of your driveway, and realize that many of the homes you see belong to people who lost a good chunk of their money because of your failure, it's very, very painful. This failure was the most cutting and damaging failure of my career. It was bad enough to have failed, but to have failed in front of your entire country was hard to take.

I've written about this failure in the past, but I am talking about it here in much starker terms than I have used before. As Canadians, we don't often write about our personal failures. I guess we still have some of the British attitude of keeping a stiff upper lip. My parents were from a generation who lived through the Second World War, and I (and so many of my generation of Canadians) was taught to never complain about our situation, since we were "lucky" not to have been involved in a world war. Good point. In the face of failure or great difficulty, we would, as the British say, "keep calm and carry on." This is an important tenet to follow in the face of adversity. Do not lose your grip.

o o o

What came out of that failure? What were the lessons learned from the decisions OpenText took to turn away from the internet and to focus on intranets?

As the first part of this chapter discusses, before OpenText ran out of money, we needed to identify something more appropriate to our skill set-up in Canada. We took the standard internet search technology, repurposed it, and used in a way that was truly innovative and, most importantly, was not being used by as many competitors. In retrospect, it looks completely sensible and the obvious thing to do. But at the time it was a huge, risk-laden (yet strategic) decision that would eventually lead to the creation of the largest software company in Canadian history, today's OpenText. Which brings up something that I have learned about Canadians, their inventiveness, and their risk tolerance.

In my experience, venture capitalists regularly report that Canadians come to the marketplace with ideas that are just as good their American counterparts' ideas. However, we aim too low. We ask for $10 million to own the Canadian market or to own a niche in the global market, whereas the Americans always ask for $100 million because they want to dominate the world. Venture capitalists repeatedly comment on the cultural differences between Canadian and American entrepreneurs. Canadians are very practical people. We generally do enough to get results, but we are not maniacal about it. Our attitude is that for us to win, others don't have to lose. Canadians are concerned about everyone winning. In other cultures, though, for them to win, not only do they have to win, but also their competitors have to lose.

Part of the motivation for Governor General David Johnston to create the Governor General's Innovation Awards was to begin the celebration of excellence in innovation, and to encourage Canadians to take more chances, to aim higher. As part of this awareness effort, David Johnston and I have written a book on the history of Canadian innovation called *Ingenious* to catalogue the many successes that the people of this country have achieved.

OpenText created a very successful business by turning to business-to-business applications behind corporations' firewalls, which was less competitive but no less lucrative or compelling than the field we moved out of. OpenText was far more defensible and far more profitable in the long run. However, if we had not failed with the first strategy, re-evaluated and pivoted when we did, we never would have realized this success.

This is, I hope, the single most important learning that comes from my story: no matter what happens, you must stick it out, adapt, evaluate, and pivot with confidence because, on your 131st try, it's going to work.

Successful entrepreneurs are actually serial failures. Thomas Edison is perhaps the greatest serial failure in the modern era — but we don't think of him that way. We think of him as a successful entrepreneur and as a successful inventor. In fact, he failed time after time. He used his failures as opportunities to learn what didn't work.

Failures are part of the journey toward success. As Edison and so many other innovators have proven, to fail is not the worst thing that can happen. But Canadians are myopic about it. Canadian companies such as Nortel and Research in Motion have had very public collapses. So have many others. The reality is that failure is going on all the time, and it is part of the innovation process.

We at OpenText were fortunate that we did change direction in the nick of time. OpenText was able to take some of its resources, some of its money, and the technologies that it had built, repurpose them to a different market and a different application, and become a great success. Ironically, this failure was only temporary, since the pivot to the intranet and the deep Web proved to be the right decision; that would not become apparent, though, until many years later. OpenText began to grow in its new space and went on to be the market leader in designing and building software (and the cloud) for

The OpenText headquarters.

corporate intranets. Today, OpenText is one of the largest software and cloud companies in the world. It never left its home in Waterloo. Today its headquarters for over twelve thousand people is located just a few steps away from where the Open Text Index was created in 1995.

o o o

OpenText's consolation prize was that it went into the business-to-business sector. In this sector, the market capitalizations are much smaller. However, they're much more stable. Business-to-business is based on long-term — decades-long — supply-chain relationships. There was far greater risk in the consumer market — fickle, fashion-and-trend-conscious as it is. Take BlackBerry or Samsung — they've both been punished by the retail market. Nokia at one time was the most powerful mobile phone company in the world, but waited too long to change. It was sold to Microsoft, and then disappeared.

One day we may write about Google or Apple falling by the wayside. In the consumer world, if you are at the top, you are king; however, there can only be one king, and usually for a very short time. OpenText is now more of a company like IBM, or SAP, or Magna, that doesn't aspire to the crown but also doesn't go down to defeat. Magna is a good example. It's a Canadian company that provides products to its industrial customers — mainly automobile companies. Magna itself is not a retail brand, and so its share price trades at a lower multiple compared to retailers. That is a very Canadian trait — we tend to be cautious and take the safer road. It may not be dramatic, but it is more dependable. OpenText became a much more reliable, long-term, stable company by taking the risk of pivoting from internet searching for the general public to focusing on intranets and document management for businesses. It quit the entire digital category of consumers and social media.

Regarding innovation: I have studied it, have seen it during my travels across the country, and I have written a lot about it. I think it is worthwhile to say that Canadians are actually quite good at innovation and coming up with new ideas. However, they underestimate themselves. When I make these comments I am not talking about any sort of inferiority, I'm talking about our attitude and our cultural bias against risk-taking.

When we do something in Canada, we do it with high quality — just as well as anybody in the world. Therefore, I think it is important that when we speak about Canadian culture, we speak about this tendency for Canadians to underestimate themselves. Canadians have succeeded at creating, innovating, and producing products and services for the global market. To use a baseball analogy: Canadians don't swing for a home run; we're very happy to get to first base. You will argue that many baseball players have had Hall of Fame careers by getting on base all the time. It's not as flashy as being a big home run hitter, but big home run hitters also strike out a lot.

There is no question there is a Canadian attitude that favours choosing lower returns for lower risk.

If you ask me, "Are Canadians afraid to fail?" I would answer that, because we are a northern people living in a vast country, we have learned that taking a risk in the middle of winter and failing lands you in big trouble. You can even die. In many other countries where the climate is more moderate and the distance between points is much less, a failure does not readily lead to a life-or-death situation. We are a product of our geography and our climate. In many situations, the consequences of our failures have been greater. I think this has had a big impact on our fabric as a country. To put it bluntly, our history in Canada has been that if you were not cautious, you removed yourself from the gene pool.

As Canada's Indigenous peoples and the early settlers knew and learned, often the hard way, being cautious and conservative is necessary. These attitudes became rooted in our history and our culture across the country. While working together with the governor general on our book, *Ingenious*, I learned about many Canadian inventions and ideas that were developed to deal with our harsh weather or our vast geography. We can't escape these elements — they are a constant in Canada.

One of the unique things about the city of Waterloo, Ontario, is that it is a place where you can regularly fail: it is treated like a badge of honour. I believe that this culture of accepting failure is one of the main reasons there are so many entrepreneurs in the city. What I have learned here is that failing, and being open and honest about it, matter. There is an old saying: *He who makes no mistakes makes nothing.* This is very important. You cannot — repeat, *cannot* — have innovation without a long history of failure. Another valid cliché: *People don't learn from success — people learn from failure.*

It's so important that we capture this stigma-of-failure narrative, understand its effects, and change it.

o o o

And yet, as I have mentioned, there is a stigma in Canada around failure.

Failure needs the perspective of time to determine what it really means. This is an important point, since there is always pressure to be the "hare" instead of the "tortoise." There is also a lot of pressure to wait "just one more month" when things are going downhill, and to be consistent with a strategy instead of admitting defeat. When I look back at these difficult moments of decision with OpenText, perhaps the most important thing about the decisions was not whether they were the right ones but rather how they were made. We made our decisions together as a group. We took the time to be careful, to examine all of our alternatives, to obtain the best evidence that we could, and to make sure that we had a consensus before we admitted failure and moved on. We were prepared to succeed or fail together as a team. That failure brought us together, and we were determined to succeed in our pivot. OpenText became the largest software company in Canadian history and one of the largest software companies in the world by sticking together and being patient in the face of failure.

The fact that you failed leads you to reflect on what happened and what went wrong. It causes you to take something apart, put it back together again, and figure out why it failed. That's why failure is such an important step in improvement.

Remember, when one door closes, another one opens.

4

DAVID T. MCNAB

*Professor in Indigenous Thought and Canadian Studies,
York University*

The Failure of Canada's Indigenous Policies in the Context of Truth and Reconciliation in 2017

I met David through the gang at the Canadian Council for Aboriginal Business, who spoke very, very highly of him. Professor McNab[1] is a Métis historian whose work has reached many people on many different levels. An expert in Indigenous treaties, land, and resource issues in Canada, he received his Ph.D. from the University of Lancaster in 1978, and was recently made a Fellow of the Royal Society of Canada. David is that rare thing: a scholar who excels in both academia and public service. He applies his abilities to solving practical problems and brings great benefit to Canada. Frankly, he is my type of human being — one who rolls up his sleeves and helps, and tries to make a difference through his contribution. I am thrilled that David wanted to contribute to this project.

— Alex Benay

The federal government has been a failure in the development of its Indigenous policies since 1867.[2] Canada was bequeathed British imperial Indigenous policies that were both colonial and racist. It has always denied Indigenous rights both in an international context and within the Canadian nation state. This consistent denial of Indigenous people, their voices, and their rights is in spite of the fact that such rights are part and parcel of Canada's Constitution (1982), especially Section 35, which states unequivocally that Indigenous people are recognized as "Indian, Inuit and Métis." Their Indigenous land and treaty rights have been reaffirmed by the Supreme Court of Canada since the Calder case in 1973.[3] At the same time, federal legislation is still on the books — the Indian Act (since 1876) — which is also both racist and colonial, and which reflects those policies. Provincial government policies are also reflective of the policies of the federal government and the Indian Act. All of this history is in spite of the many initiatives taken by Indigenous peoples in Canada to change these policies and the day-to-day processes of the federal government, as well as to resist their implementation.

The issue is a fundamental one, that of Indigenous sovereignty. The same is true on the international stage. The federal government has until recently denied Indigenous rights in Canada. Indigenous coalition representatives say they believe the opposition by the world's big powers was largely driven by concern over the potential loss of state control over how natural resources, like oil, gas, and timber, are exploited.[4] Nor should this denial of Indigenous rights be at all surprising, given the policies initiated as process within the place known as Canada since the early seventeenth century by European empires — the French (to 1763), the British (to 1867), and then the Canadian empires (since 1867).

What is this policy and how is it formulated in Indigenous contexts? Policy is multidimensional and multifaceted, as J.W. Cell has noted, "being something rather less fixed, something rather more historical." At any moment in time there is "not so much policy as policy formation, an unsettled and changing set of responses by government to the continual interaction among men [and women], forces, ideas, and institutions."[5] As such, the Indigenous policies of Canada promulgated by the nation state have their origin in the history of the treaty-making process of Canada. Canadian Indigenous policies owe their development to the many misrepresentations of how Europeans have viewed and represented Indigenous peoples, including First Nations, Métis, and Inuit peoples. However, First Nations, Inuit, or Métis have no need for such policies.[6] They know who they are. The land "is the boss" and the land tells them.[7]

Creations of the nation state of Canada, these policies are an aberration of the Covenant Chain of Silver and especially the Two Row Wampum. The latter represented the relationship between the Dutch, English, and French imperial governments and the Indigenous Nations: namely, peace, mutual respect, and trust. Initially given by the English to the Haudenosaunee to cement the treaty entered into at Albany in 1664, its components were known as least as early as the early seventeenth century when the very first Treaties of Peace and Friendship were entered into between the French and the Mi'kmaq Nations in Acadia — present-day Atlantic Canada. The Covenant Chain literally means "to link one's arms together," and signifies a nation-to-nation relationship.[8]

The significance of the Covenant Chain of Silver as a basis for the treaty-making process and Indigenous policies cannot be underestimated in terms of land and sovereignty. Sir William Johnson, the English Crown's imperial appointee to the Indian Department in 1755, highlighted its magnitude in 1764 when he wrote:

> Tis true that when a Nation find themselves pushed, their Alliances broken, and themselves tired of a War, they are verry [sic] apt to say many civil things, and make any Submissions which are not agreable [agreeable] to their intentions, but are said meerly [sic] to please those with whom they transact Affairs as they know they cannot enforce the observance of them, but you may be assured that none of the Six nations, Western Nations [including

the Western Confederacy, also now known as the
Anishinabe Nations] &ca. ever declared themselves
to be Subjects, or will ever consider themselves in
that light whilst they have any Men, or an Open
Country to retire to, the very Idea of subjection
would fill them with horror.[9]

This statement by Johnson links the basis of this process with
one of the early views also integral to Canada's Indigenous policies
— the notion that Indigenous peoples were clearly not subject to an
empire or a nation state. This notion of "subjects" of the Crown, rather
than sovereign First Nations, was a direct result of the ideology of
European empires, notably the French and the British empires, which
sought to control and dominate the natural world of North America
and the peoples who resided there. Initially, outnumbered and without sufficient technology (the canoe was one of the primary modes
of resistance) to dominate Indigenous peoples, the French and then
the British empires, recognizing Indigenous peoples as nations, sought
out and entered into Treaties of Peace and Friendship.[10]

One of the first unilateral statements of British imperial policy toward the First Nations was the promulgation of the Royal
Proclamation of 1763, which was partly a response to the Anishinabe
and Seneca resistance movements earlier that year. Owing much to
the treaty-making process, the Royal Proclamation was an English
imperial document, among other things, that recognized and reaffirmed that the "Indian territory" was to be their "absolute property." It established English imperial rules regarding the treaty-making
process under the Covenant Chain as well as recognizing the significance of the sovereignty of Indigenous trade and trading.[11] It would
be reaffirmed one year later in a grand council of Nations at Niagara
in 1764, and in subsequent treaties. The following is a First Nations'
perspective on this Treaty of Niagara:

> While the treaties are like stones marking a spot in
> time, the relationship between the Nations is like
> two equals, respecting each of their differences but
> supporting each other for a common position on
> peace, order and justice for all. The brotherhood
> created by the Twenty Four Nations Belt represents
> a relationship of both sharing and respect. The shar-
> ing is reciprocal: as the First Nations shared land
> and the knowledge in the past, now that situation
> is reversed, the generosity of spirit and action is ex-
> pected to continue. The respect is also reciprocal:
> respect for each other's rights, existence, laws and
> vision of the future.[12]

This Indigenous view of treaties and treaty-making has not changed before, or after, 1764. Herein lie the failures of Canadian Indigenous policies, which were constructed upon different principles.

By the late eighteenth century, the balance of military power was beginning to shift. Increasingly the treaties came to be seen, not as sharing the land and natural resources, but as land "surrenders." Indigenous peoples were also being seen by non-Indigenous people as "savages" and as subjects of the imperial crowns. These promises were not inconsequential at a time when the English imperial foothold on the North American continent was at best precarious. Sir William Johnson died in 1774, and after that, things began to fall apart. The solemn promises of the Crown, made with wampum belts, to which were attached the writings of the participants, were forgotten by the Indian Department by the 1790s.[13]

After the disaster for Indigenous people wrought by the American War of Independence (1775–83), the new colony of Upper Canada became of enormous significance to the British imperial government as a settlement colony for its lands and resources. It was also very important, for its strategic military advantage, to protect this English colony from the United States by creating a sovereign Indigenous buffer state, effectively a military zone between the new American republic and Upper Canada. The land was also seen to be, in the long term, of great value as a place for the second British Empire to promote commercial agricultural settlement and colonization. Lieutenant-Governor John Graves Simcoe's plan for Upper Canada (1791) outlined the following:

> There are but a very few Indians who inhabit within it, the greater part of the soil has been purchased & the whole ought to be before it will become of value, as the Indians will not want for suggestions to inhance [sic] its price. I consider the Country to be of immense value, whether it be regarded to its immediate advantages, the future prospect of Advantage, or the probable grounds for supposing it will remain the most important foreign possession of Great Britain.[14]

Based ostensibly on the Royal Proclamation of 1763 and a complete misrepresentation — as well as a misconstruction — of the treaty-making process, Simcoe's land policy was instead premised on a number of land surrenders in Upper Canada, which occurred from the 1790s to the Confederation of Canada (1867), and in the numbered treaties thereafter in the Canadian West and North. These treaties were not entered into without resistance by First Nations' citizens. The embodiment of the Two Row Wampum and the ostensible promise of protection of subject peoples soon became a land grab filled with corruption and land speculation, such as were similar cases at the time in the United States.

The War of 1812 was a "turntable" for Canadian Indigenous policies.[15] Although fighting for the British in that war, First Nations, after the war, were no longer needed as military allies. The result was that a boundary was agreed upon by the United States and Britain, but the survey was never completed through the Great Lakes, and thus was never confirmed by executive authority of the American or the British imperial government, or Canada after 1867. This boundary split First Nations' territories, literally carved them up by the survey of the international boundary and, more seriously, effectively took away jurisdiction and governance (but not sovereignty) from First Nations.[16]

Relocation and attempted extinguishment of Indigenous title and reserve lands in Upper Canada followed the War of 1812. Gradually, a full-blown British imperial policy of "civilization" (the origins of "cultural genocide" in residential schools promulgated by the Indian Act of 1876) was established by the 1820s.[17] Many more requests came for land surrenders from the English Crown, but the existing land policies were not followed. Most of the monies were not deposited to trust fund accounts, and more lands were taken rather than shared, as outlined in the written treaty documents. There were also Indigenous oral stories and traditions from wampum belts about the ceremonies from treaty-making.[18] Even without the overt use of force, the British imperial government sometimes used another process, a process based on non-consultation and non-consent — in violation of the spirit and intent of the Royal Proclamation and the treaties — to achieve the same end of extinguishment.

Beginning in the 1820s, with increasing British immigration, agricultural settlement by white settlers began on a large scale. Requests came for more land surrenders. In the 1820s, under governors Darling, Colborne, and then Sir Francis Bond Head, the British imperial government embarked on its policy of "civilization" in Upper Canada, since it had already been deemed a success by English imperial officials in Lower Canada. While this policy was primarily designed with a hard edge to assimilate, purposefully, Indigenous people, it also promised monies for education and training and economic opportunities — for example, commercial agricultural opportunities — for the citizens of the First Nations. Another less desirable approach of the policy was a conscious plan, developed by the lieutenant governor, Sir Francis Bond Head, in the mid-1830s, to remove Indigenous people from their territories. The idea was to remove and to centralize Indigenous people into two geographical areas, primarily Manitoulin Island and Walpole Island. Once centralized on islands, they could better be, according to the government view, "civilized" and then assimilated. This was strenuously resisted by the First Nations. The only group that became enfranchised as "white people" by 1880 under this policy were the Wyandot of Anderdon, south of present-day Windsor.

This so-called "civilization" policy was not predicated on the surrender of Indian lands, at least initially, to pay for the policy. Rather, the monies for it would come from a general parliamentary grant from the Crown. Subsequently, the policy became one of forced assimilation, which was financed crudely by selling First Nations' lands and using the funds raised to pay for their own assimilation. This policy was anathema to First Nations. By 1840 it had already failed at Coldwater and other places. Yet the Indian Department, confronted by wholesale squatting and trespass on First Nation lands, continued to implement it by the taking of land surrenders. This policy was codified in the late 1850s in the first Indian Act of 1857. The import of this new approach to civilization was not lost on the First Nations leadership, who rejected it and bluntly stated that it was an attempt "to break them to pieces."[19]

Partially responding to the attempted encroachments and the alienation of the "Indian Territory," the English imperial government took action, flowing from the Royal Proclamation, to protect parts of the Indian Territory. In 1839 it passed legislation to protect Crown lands, especially the Indian Territory, which had been the subject of considerable concern because of trespass, squatting by non-Indigenous people, illegal land use (for example, the taking of timber from Indian lands), and outright fraud. However, this legislation proved not to be strong enough in the decade following its passage, and the illegal taking or other depredations, such as trespassing or the taking of natural resources, on Indigenous Territories, unceded lands, and reserves, continued. On August 10, 1850, the government of the Province of Canada passed further legislation, an act for the "protection" of the "property occupied or enjoyed" by Indigenous people in Upper Canada "from Trespass and Injury." This legislation strengthened the provisions of the 1839 Act, but the legislation still appears not to have been effective, since the squatting and dispossession continued unabated. There was little or no enforcement of this law. However, Indigenous people did not become citizens of Canada until John Diefenbaker's first Canadian Bill of Rights was established in 1960.[20]

In 1861 Herman Merivale, an astute British imperial commentator and a consummate bureaucrat, observed that British imperial Indigenous policy had been a failure. His commentary could well be a description of Canada's Indigenous policy more than 156 years later:

> The subject, in short, is one which has been dealt with by perpetual compromises between principle and immediate exigency. Such compromises are incidental to constitutional government. We are accustomed to them: there is something in them congenial to our national character, as well as accommodated to our institutions; and on the whole, we may reasonably doubt whether the world is not better managed by means of them than through the severe application of principles. But, unfortunately, in the special subject before us, the uncertainty created by such compromises is a greater evil than errors of principle.[21]

Merivale described the vacillation of Canada's Indigenous policy (or policies) as flowing from the perpetual compromises between principle and immediate exigency. This is a significant observation about the failure of Canada's Indigenous policies. These policies have created great uncertainty and extreme frustration with the failure of the Crown to uphold the Covenant Chain of Silver, the sharing of the land and natural resources, and the concomitant solemn treaty promises. This situation helps to explain the failure of Canada's Indigenous policies.[22]

One of the primary events that eroded Indigenous rights was the Confederation of Canada in 1867. In that year the British North America Act was passed by the British imperial government and Indigenous Nations effectively appeared to have lost recognition and respect for their rights of citizenship and sovereignty over their governance and over their lands and waters. This situation is changing gradually because of Supreme Court of Canada rulings since 1973. The new federal government assumed responsibility for "Indians, and Lands reserved for the Indians" by Section 91 (24), subject to any liabilities, which the government of the Province of Canada had, to Indigenous Nations.[23] This legislation allowed the federal government to pass the first consolidated Indian Act, thereby establishing a colonial relationship of the federal government to First Nations. It also stated that the provinces had control over all other lands within the boundaries of each province (Section 109). However, this imperial statute was subject to any outstanding interests, including reserve lands as well as the Indigenous Territory, much of which was still unceded, and neither the interests nor the lands were specified. If the lands were not referred to, then it was subsequently assumed that Indigenous Nations' land rights did not exist. Originally, the Confederation of Canada was conceived of, and was supposed to have been, a treaty among the founding nations of Canada, including all of the Indigenous Nations, based on the Two Row Wampum. But it soon became a means of carrying forward the policy of extinguishment, including the surrender or relinquishment of the Indian Territory, as well as the implementation in the late nineteenth century of the residential school system and its horrific cultural and sexual abuses, which lasted well into the 1980s. The last residential school only closed in 1996.[24]

Conflict, rooted in imperial and colonial aspirations, as well as cultural disparateness, grew apace as Canadian Indigenous policies continued to develop within regional frameworks. There were treaties of peace and friendship in Atlantic Canada; no treaties in Quebec or the Far North; land loss treaties in Upper Canada, which became a model of the subsequent numbered treaties (1871–1930) in Canada's West and North. The oft-stated principle of "protection," through imperial trusteeship (which was never fulfilled), gave way to assimilation, and with it came European scientific racism as part and parcel of Canada's Indigenous policies.[25]

The dichotomy between the two principles and the discrepancies over the negotiation of the so-called numbered treaties in the late nineteenth century was a clear example of the failure of this process. The divergence, not only in the treaty-making process in the 1870s, showed also the weaknesses of Canada's regional Indigenous policies at that time and thereafter. Mawedopenais, a Mide Chief of the Ojibwa, spoke to the Crown's commissioner and chief negotiator, Alexander Morris, at the Treaty #3 negotiations in October 1873. As a spokesperson for the Rainy Lake and Rainy River people in this treaty-making process, he was clear on the position of Indigenous Nations and the title to their lands:

> I lay before you our opinions. Our hands are poor but our heads are rich, and it is riches that we ask so that we may be able to support our families as long as the sun rises and the water runs. Morris replied, disingenuously, indicating that he did not understand what Indigenous title and the Treaty-making process meant for the Indigenous Nations: "I am

very sorry; you know it takes two to make a bargain; you are agreed on the one side, and I for the Queen's Government on the other. I have to go away and report that I have to go without making terms with you. I doubt if the Commissioners will be sent again to assemble this nation."[26]

This threat, implying the government approach of "divide and conquer," was not, as may be expected, well received by the Anishnabek Nation. Treaty #3 was eventually negotiated and signed, but not on the basis of the treaty document or as understood by Morris. He did not believe, as many people do to this day, that the Indigenous Nations had been ready to share in the treaty-making process with the riches in their heads. There was no balance in the "bargain" before or after the treaty was signed. Morris and the federal government took too much away from the life of the Ojibwa.[27] It has continued to do so here and elsewhere in Canada. From this unequal perspective, these negotiations were not successful. It was not the only treaty that could be characterized in this fashion. Yet the treaty issues do not die. They live within a circle of time for later generations. Indigenous people in Canada resist and they never forget.

This non-consultative treaty-making approach became central and pivotal to the development of the top-down policy approach of the federal government inherent in the Indian Act of 1876 and its successors. Moreover, except for one substantial revision to that legislation in 1951, the Indian Act remains on the books today as a cornerstone of Canada's Indigenous policies. Why is that legislation racist? It is not consultative or community-based. It is the federal government that decides who is an Indigenous person or not under the registration process of that act (which is still based on a racist approach using blood quantum); non-status and Métis persons are not and cannot be registered under it, notwithstanding that they are recognized in Canada's Constitution as Indigenous peoples. In fact, it was not until the 1930s that the Inuit of Canada's North were recognized through a court ruling as having equal status as "Indians." Yet there is no Inuit Act today. Canadian Indigenous policies have been and still are in complete disarray.

More than one hundred years later there is still a wide cultural gulf in the treaty-making process. This has intensified and has led to the abrogation of Indigenous title and treaty rights and to the events of the summer of 1990 at Oka. But these events were broader than the events at Kahnesatake and Kahnawake. Similar situations also occurred in northern Ontario and in British Columbia. In 1995, the denial of Indigenous rights led to the death of Dudley George by an OPP officer at Ipperwash.[28] They are still occurring in Ontario at Caledonia with the original Six Nations reserve. In each instance Indigenous rights were denied by the Canadian nation state. At Oka (as also happened in 1869–70 and 1885 during the Métis resistance movements) the Canadian Armed Forces went in to enforce the denial of Indigenous rights.

What accounts for these differences in Canadian Indigenous policies and the reality of Indigenous existence in Canada? The answer lies in the disparate histories of Indigenous and non-Indigenous people in Canada. The primary objective of Indigenous people is a spiritual, peaceful one: to protect the land — Mother Earth — and the waters of Turtle Island. This is a sacred trust, a trust to protect the land. The continuity and integrity of their lands are important to their survival as Indigenous people. Generations of Indigenous Nations' members have used the land and have shared in its bounty and its uses. Moreover, they will continue to use this land and teach their children about the Creator and the Land. So this relationship is all-important. They owe their very survival to it. It is both simple and profound. The events of the summer of 1990 at Oka and elsewhere across Canada occurred in our time at the initiative of Indigenous people to protect their lands and waters. To do this, they had no choice but to resist those who wished to destroy the land and themselves. Not to resist would mean their destruction, as well as the destruction of their children and grandchildren. It would have meant the end of their cultures and survival as Indigenous people. They will continue to protect their lands and waters.

Canada's Indigenous policies are very gradually becoming irrelevant. In the twenty-first century we are witnessing profound structural changes in the history of the world. The world of nineteenth-century and twentieth-century European and American imperialism is over. Decolonization is continuing apace. This process has been characterized by forces of both construction and destruction. In Canada, to provide one example, Indigenous peoples are again reaffirming their sovereignty and their inherent right to governance through diverse approaches and means. Their lands are ever so slowly being recovered; if not always respected. Indigenous title is beginning to be understood and recognized. One watershed in the twentieth century is the Calder case of 1973. This was the Supreme Court of Canada decision that found that Indigenous title and rights did exist in the white justice system of Canada. It opened the legal doors for the prosecution of Indigenous title and rights cases in Canada. Calder was followed by many constructive Supreme Court of Canada decisions that reaffirmed Indigenous title and land rights, and that affirmed treaties, including Guerin, Simon, Sioui, Sparrow, Delgamuukw (1997) and Marshall (1999), Manitoba Métis (2013), and Daniels (2016), to name but a few.[29]

Calder also opened the door to the new land claims policies of the federal government in 1974, which has operated to bring about an undermining of Canada's historic Indigenous policies, and a replacement of those policies with various forms of Indigenous governance. This has in turn led to the creation of the Territory of Nunavut in 1999, and then to the first modern treaty — the Nishga Treaty — in British Columbia in 2003.

What is the basis of this new Indigenous policy, which is embodied in the land claims processes? There are no Indigenous land "claims," only land rights. There are only Indigenous title and treaty rights to the land, and these must be protected. Land grievances flow from the treaty-making process. And these areas are not restricted to reserves, places of special and specific protection. As strategic areas of land for Indigenous resource use, they were a major consideration for both the Crown and Indigenous people, but reserves and territories form antithetical concepts of land holding that still intensify conflicts over treaty areas and reserves.

The Indian Claims Commission was established after the events of the summer of 1990 at Oka and elsewhere. That Commission can report and publish its findings and recommendations to the federal Cabinet. But the federal government does not have to implement its recommendations. The Commission has continued. What is a land claim? It is a statement of the land rights of an Indigenous nation, which contains reference to a specific geographical area. It is a claim based on whether the Indigenous resource users of the land in question have ever decided to enter into a treaty for it under the rules set out in the Royal Proclamation of 1763. A claim is not a court action; it is policy, not litigation. It is a policy (actually two policies — one for comprehensive claims and the other for specific ones) and a program of the federal government administered by the Department of Indian and Northern Affairs. The government's purpose is to extinguish "claims" and thus effectively deny Indigenous rights. The process within government is restricted and, insofar as a claimant decides on litigation, the file wends its way through the federal bureaucracy to the Department of Justice, where the claimant is seen from the policy and program point of view to have withdrawn the claim. The bureaucrats then arbitrarily close the claim and place it on a dusty shelf along with others, like a mere curio on a Victorian collector's shelf. It is literally another form of "captured heritage." In the late 1980s one senior federal bureaucrat even used this very image of Victorian curios to arrogantly describe this situation with paternalistic condescension.[30]

Land rights are significant if they can, through substantive settlement agreements, add to the land and the economic base of First Nation communities. Indian reserves were initially strategic economic areas that were exempted from the treaty-making process. Later they were transformed and designed by British imperial policy-makers as special areas of "civilization" with the specific objective of assimilating First Nations. Soon they became mere "halfway houses," which were to be appropriated whenever they were needed for the purposes of the Crown or for other non-Indigenous uses. Thus, a form of British imperial trusteeship gave way en route to a path leading to gradual, and then it was hoped, complete, assimilation. This form of "cultural genocide" is a vestige of a pure colonial relationship. Even this misguided sense of Indigenous "claims," distorted by twentieth-century lenses, has been rendered illegitimate over time by the alienation of land and labour of Indigenous people. Land "surrenders" as well as the loss of the commons for natural resource exploitation by governments and by private interests assisted the process. What was seen to be legitimate was rendered both unlawful and unfair from an Indigenous perspective. What cannot be seen by governments cannot be destroyed. The Temagami case, and in particular the building of the Red Squirrel Road extension in 1988–89, is a prime, but not a solitary, illustration.

The repatriation of Canada's Constitution in 1982 single-handedly initiated by Prime Minister Pierre Elliott Trudeau was a watershed of Indigenous rights in Canada.[31] The Constitution was critical to recognizing Indigenous rights in what became a legal context. As a result of many years of Indigenous resistance movements, the Ontario provincial government, on October 2, 1990, recognized the inherent right of Indigenous governance. The federal government followed suit a few years later, but it still remains outside of the Canadian Constitution (1982) today. The Meech Lake Accord of 1987, the epitome of the old British imperial centralist model of Confederation, stylishly referred to as "executive federalism," failed to be ratified in

1990. This was a clear Constitutional victory by Indigenous people. They are in the Canadian Constitution. Although the Charlottetown Accord of 1992 was also a failure, the inherent right of Indigenous people to Indigenous governance, as well as their Indigenous title and land rights, have since been recognized and reaffirmed, but not yet in our Constitution.

This reversal of British imperial policy was only altered in 1982 when Canada's Constitution was patriated and renewed; "existing Indigenous and treaty rights" were admitted as part of it, but not Indigenous governance or sovereignty. Today the larger business of the Constitution and the treaty-making process through various land claims policies still remains incomplete and unfulfilled. It is currently being defined using Canada's Constitution on an issue-by-issue basis by the Supreme Court of Canada. This concept of Indigenous oral history and traditions has recently been reaffirmed by the Royal Commission on Indigenous Peoples (1996), but never implemented.

Most notably, in 1997 the Supreme Court of Canada ruled, in part, on this issue, in the case of *Delgamuukw v. British Columbia*, also identified as the Gitksan and Wet'suwet'en comprehensive "claim." That legal ruling stated that oral traditions are "not simply a detached recounting of factual events but, rather, are 'facts enmeshed in the stories of a lifetime.'" Moreover, they are "rooted in particular locations, making reference to particular families and communities." As a result, Indigenous oral history is in fact "many histories, each characterized in part by how a people see themselves, how they define their identity in relation to their environment, and how they express their uniqueness as a people." The Supreme Court stated that the "laws of evidence" in the Canadian justice system must accommodate Indigenous oral history and traditions such that it "be placed on an equal footing with the types of historical evidence that courts are familiar with, which largely consists of historical documents. This is a long-standing practice in the interpretation of treaties between the

Crown and Indigenous peoples." Not to recognize and accept this history as an equally valid way of viewing the past is to deny Indigenous people and their land rights.

Many of Canada's modern Indigenous land policies are still viewed one-dimensionally as primarily assimilative, as a form of forced and directed cultural change, which, in the case of residential schools, became "cultural genocide." This has been seen as originating in the nineteenth century and culminating in Jean Chrétien's White Paper of 1969. Canada's Indigenous policies have developed gradually as two primary components that are diametrically opposed to each other. They become built-in obstacles. The first is that the federal government has been largely indifferent to Indigenous title and land rights, taking a legalistic approach overall, only acting when it is forced to do so by Canada's courts. The second component is that the provinces continue to use their hegemony, through legislation and regulations, over lands and natural resources in self-serving ways. Canada's Indigenous policies since 1867 have been an artificial creation, both negative and destructive, for Indigenous people and their relationship to the rest of the country.[32]

Canada's Indigenous policies, formulated through colonialism, through a long process of denial, have created institutional racism and corresponding resistance movements that culminated in violence and death. The events of the summer of 1990 at Oka have not been erased. The initiative for change in Indigenous history has always come from the First Nations. Federal and provincial government policies have always been characterized by reaction, crisis management, and denial. Encountering policy words with no substance and a benign, passive policy, Indigenous nations have always chosen to act. They had no choice but to act to resist these policies — if they wished to survive. Indigenous people will continue to resist and survive.

In the early twenty-first century the prominent issues arising from Canada's Indigenous policies remain outstanding and unresolved. In this sense, Canada's Indigenous policies have been a wholesale failure in the face of the resistance to them by Indigenous citizens. And they have led directly to the consistent denial by the federal government of Indigenous rights in Canada and on the international stage.

There is hope. On June 2, 2015, after six years of research, (*Final Report* delivered on December 15, 2015), the Truth and Reconciliation Commission issued its 383-page executive summary report, *Honouring the Truth, Reconciling for the Future,* making ninety-four recommendations in the form of "calls to action." The report flowed from some of the residential schools issues that had been raised since at least the last quarter of the nineteenth century. This report was published in six volumes late in 2015, and it spoke to the multidimensional issues of the seven generations of Indigenous people, and others within First Nations communities who have been affected by residential schools from 1885 to 1996 and thereafter. The report's findings were stated at the outset:

> For over a century, the central goals of Canada's Aboriginal policy were to eliminate Aboriginal governments; ignore Aboriginal rights; terminate the Treaties; and, through a process of assimilation, cause Aboriginal peoples to cease to exist as distinct legal, social, cultural, religious, and racial entities in Canada. The establishment and operation of residential schools were a central element of this policy, which can best be described as "cultural genocide."[33]

This history was built on the back of the system of colonialism and racism initiated by the British Empire and then continued by the Canadian Empire (embodied politically in the nation state) since 1867. This history is fully developed, documented, and shown in the "History" section of the *Final Report.*

On May 29, 2015, before the report was formally given, the Chief Justice of Canada, Beverley McLachlin (the highest ranking Canadian official to have ever used the term "genocide") stated in a public lecture that the "most glaring blemish on the Canadian historic record relates to our treatment of First Nations that lived here at the time of colonization," and that Canada had developed an "ethos of exclusion and cultural annihilation." She spoke as well of "cultural genocide."[34] The report defined what it meant by the term "cultural genocide." It is the

> destruction of those structures and practices that allow the group to continue as a group. States that engage in cultural genocide set out to destroy the political and social institutions of the targeted group. Land is seized, and populations are forcibly transferred and their movement is restricted. Languages are banned. Spiritual leaders are persecuted, spiritual practices are forbidden, and objects of spiritual value are confiscated and destroyed. And, most significantly to the issue at hand, families are disrupted to prevent the transmission of cultural values and identity from one generation to the next. [35]

In its dealings with Aboriginal people, Canada did all these things.[36]

The first call to action, and one that is central to all ninety-four recommendations, was for Canada to endorse the United Nations Declaration of Human Rights. The current government accepted the *Final Report* of the Truth and Reconciliation Commission when it was released on December 15, 2015. Some of these calls to action have already been adopted. However, it will take seven generations to bring about change caused by the failure of Canada's Indigenous policies.

Notes

1. David T. McNab is Professor of Indigenous Thought and Canadian Studies, Faculty of Liberal Arts and Professional Studies, York University, Toronto, Ontario. He was made a Fellow of the Royal Society of Canada in September 2017. An earlier version of this chapter was originally published as "A Brief History of the Denial of Indigenous Rights in Canada," in *A History of Human Rights in Canada*, ed. Janet Miron (Toronto: Scholarly Publishing, 2009), 99–115. It has since been updated and thoroughly revised eight years later.

2. In this chapter, I will use the modern term "Indigenous," or refer to specific groups by the name they use, and not "Indian," except for the Indian Act. Our Constitution uses "Aboriginal."

3. Calder et al. v. Attorney-General of British Columbia, [1973] S.C.R. 313, https://scc-csc.lexum.com/scc-csc/scc-csc/en/item/5113/index.do.

4. Richard Reynolds, "UN Council Approves Indigenous Rights Treaty," *ABC News Online*, June 30, 2006.

5. J.W. Cell, *British Colonial Administration in the Mid-Nineteenth Century* (New Haven: Yale University Press, 1970). See especially his introduction.

6. Ute Lischke and David T. McNab, eds., *Walking a Tightrope: Indigenous People and Their Representations* (Waterloo: Wilfrid Laurier University Press, 2005). See also, in an Ontario context, my *Circles of Time: Indigenous Land Rights and Resistance in Ontario* (Waterloo: Wilfrid Laurier University Press, 1999).

7. Gary Potts, in Bruce W. Hodgins, Ute Lischke, and David T. McNab, eds., *Blockades and Resistance: Studies in Actions of Peace and the Temagami Blockades of 1988–89* (Waterloo: Wilfrid Laurier University Press, 2003), 20–21.

8. David T. McNab (edited for Nin.Da.Waab.Jig.), *Earth, Water, Air*

and Fire: Studies in Canadian Ethnohistory (Waterloo: Wilfrid Laurier University Press, 1998), 41–44.

9. *The Papers of Sir William Johnson*, ed. Milton W. Hamilton, vol. 11 (Albany: University of the State of New York, 1953), 395–96.

10. David T. McNab, with Bruce Hodgins and S. Dale Standen, "'Black with Canoes': Indigenous Resistance and the Canoe: Diplomacy, Trade and Warfare in the Meeting Grounds of Northeastern North America, 1600–1820," in *Technology, Disease and Colonial Conquests, Sixteenth to Eighteenth Centuries: Essays Reappraising the Guns and Germs Theories*, ed. George Raudzens (Amsterdam: Brill International, 2001), 237–92; "Actions of Peace," in *Blockades and Resistance*, ed. Hodgins, Lischke, and McNab, 1–6.

11. "October 7, 1763, Royal Proclamation of 1763," in *As Long as the Sun Shines and the Water Flows, A Reader in Canadian Native Studies*, eds. Ian A.L. Getty and Antoine S. Lussier (Vancouver: University of British Columbia Press, 1983), 29–37.

12. Nin.Da.Waab.Jig. Files, Walpole Island (Bkejwanong) First Nation.

13. See my "'The land was to remain ours': The St. Anne Island Treaty of 1796 and Aboriginal Title and Rights in the Twenty-First Century," in *Native American Speakers of the Eastern Woodlands: Selected Speeches and Critical Analyses*, ed. Barbara Alice Mann (New York: Greenwood Press, 2001), 229–50.

14. *The Peter Russell Papers,* ed. E.A. Cruikshank, vol. 2, 1798–99 (Toronto: Ontario Historical Society, 1933), at October 22, 1798, 42–43, 53, 290–91.

15. Olive Patricia Dickason with David T. McNab, *Canada's First Nations, A History of Founding Peoples from Earliest Times*, 4th ed. (Toronto: Oxford University Press, 2009,) 184–93. The fifth edition was completed by the current author in 2016.

16. See my "Borders of Water and Fire: Islands as Sacred Places and as Meeting Grounds," in *Aboriginal Cultural Landscapes*, ed. Jill Oakes and Rick Riewe (Winnipeg: Aboriginal Issues Press, 2004), 35–46.

17. See Truth and Reconciliation Commission, *Canada's Residential Schools: The History, Part 1, Origins to 1939* (Montreal: McGill-Queen's University Press, 2015), 1–6.

18. There was also a Métis example of using wampum belts in such a treaty in 1818. See my "The Historical Significance of the Pierre Piche Wampum Strings of 1818," in *Tecumseh's Vision: Indigenous Sovereignty and Borders Since the War of 1812*, eds. Ute Lischke, et al. (Winnipeg: Aboriginal Issues Press, 2015), 17–27.

19. See J.S. Milloy, "The Early Indian Acts: Developmental Strategy and Constitutional Change," in *As Long as the Sun Shines and Water Flows*, eds. Getty and Lussier, 56–54; and *Indian Act Colonialism: A Century Of Dishonour, 1869–1969*, Research Paper for the National Centre for First Nations Governance, May 2008, 1–6.

20. James W. St. G. Walker, "'Race,' Rights and the Law in the Supreme Court of Canada," Historical Case Studies (Waterloo: Osgoode Society for Legal History and Wilfrid Laurier University Press, 1997), 326.

21. David T. McNab, "Herman Merivale and Colonial Office Indian Policy in the Mid-Nineteenth Century," *Canadian Journal of Native Studies* 1, no. 2 (1981): 277–302. Reprinted in *As Long as the Sun Shines and Water Flows*, eds. Getty and Lussier, 85–103.

22. For another example, see my *No Place for Fairness: Indigenous Land Rights and Policy in the Bear Island Case and Beyond* (Montreal: McGill-Queen's University Press, 2009).

23. The Métis Nation appears to be the exception. See my "The Metis Nation and the New Empire of Canada as a Place: A Perspective from Spirit Memory," unpublished manuscript.

24. See Truth and Reconciliation Commission, *Canada's Residential Schools*.

25. See my "'A Lurid Dash of Colour': Powassan's Drum and Canada's Mission, the Reverend William and Duncan Campbell Scott," in *Aboriginal Cultural Landscapes*, eds. Oakes and Riewe, 258–71.

26. See my *Circles of Time;* also see my "The Administration of Treaty #3: The Location of the Boundaries of Treaty #3 Indian Reserves in Ontario, 1873–1915," in *As Long as the Sun Shines and Water Flows,* eds. Getty and Lussier, 145–57.

27. Ibid.

28. See the Ipperwash Inquiry at www.ipperwashinquiry.ca, September 11, 2006.

29. McNab, "The Spirit of Delgamuukw and Indigenous Oral Traditions in Ontario," in *Beyond the Nass Valley: National Implications of the Supreme Court's Delgamuukw Decision,* ed. Owen Lippert (Vancouver: Fraser Institute, 2000), 273–83.

30. McNab, *Circles of Time,* 1–10.

31. See my "'A Singular Joy,' The Memorable Mindscape of Kit Tatro," in *Studies in the Literary Achievement of Louise Erdrich, Native American Writer: Fifteen Critical Essays,* ed. Brajesh Sawhney (Lampeter, UK: Edwin Mellen Press, 2008), 256–60.

32. McNab, *Circles of Time,* 187–202.

33. See Truth and Reconciliation Commission, *Canada's Residential Schools,* 1.

34. Gloria Galloway and Bill Curry, "Residential Schools Amounted to 'Cultural Genocide,' Report Says," *Globe and Mail,* June 2, 2015, accessed June 25, 2015, www.theglobeandmail.com/news/politics/residential-schools-amounted-to-cultural-genocide-says-report/article24740605/.

35. See Truth and Reconciliation Commission, *Canada's Residential Schools,* 1–6.

36. Ibid.

5

ANDY NULMAN

CEO of Play The Future, Co-Founder of the Just for Laughs Comedy Festival

My Greatest Failure

I was introduced to Andy by David Usher, the former singer of Moist who currently fights for climate change awareness and continues to stay involved in digital technologies around the country. David offered to introduce me to the creator of Just For Laughs ... I mean ... c'mon ... who wouldn't want to meet him, right? Andy came as advertised: hilarious, entrepreneurial, genuine, and to the point. His chapter on his failure in the entertainment sector is straight from the heart, and that's what I love about him — there is very little room for bullshit with Andy. I know you will enjoy his very personal contribution to this project.

— Alex Benay

This is the true story of a successful speech about the subject of failure that focused on the abysmal failure of an ambitious speech about surprise.

Got that?

Don't worry, even if you force yourself to read and reread it a few times, my opening line won't get much clearer. If nothing else, at least it's a perfect fit with the overarching theme of this book — for it is a failure.

And quite a monumental failure at that, for it repudiates the golden rule of writing, particularly the conventional wisdom of opening lines, which experts say should be crystal clear to ensure readers understand your main point, and where your story is heading.

Yet here we are, five paragraphs in, and you probably still have no idea what the hell I'm talking about. Therefore, major fail, correct?

Well, yes — as it stands right now. Major fail indeed, for the time being. However, as you are about to learn, as I am about to prove, "failure" is not a rock-solid moment frozen in time, but a metamorphous beast that fluidly changes forms and loses potency throughout its lifespan. In fact, when this story is all over, you'll discover that my opening line actually makes perfect sense.

It just has to be put into context first.

Let's start contextualizing by heading back to May 26, 2010; more specifically, to the John W. H. Bassett Theatre, tucked away in the underbelly of the Metro Toronto Convention Centre. On the afternoon of a very pleasant spring Wednesday, I was booked in that esteemed hall to give a keynote speech at the annual Canadian Marketing Association convention.

My speech's official title was listed in the event's program as "A World Without Surprise," reflecting a book I had just written about the power of surprise in business.* But in the mere twenty minutes it took me to deliver it, the speech took on new description among those now uncomfortably seated in the John W. H. Bassett Theatre audience, who unofficially retitled it "The Worst CMA Keynote of All Time."

Me, I needed even fewer words to describe it. I called it "My Greatest Failure."

Here's how it came to be so.

––––––––––

* Called *Pow! Right Between the Eyes! Profiting from the Power of Surprise!*, it's perhaps one of the longest book titles of all time, certainly the one with the most exclamation marks, and, surprisingly, still available.

A Magnum Opus for a Keynote

The story's starting point can be tracked back to mid-2009, when I was asked by my friend, Twist Image agency president Mitch Joel, to join the planning committee of the 2010 CMA event. He was the conference's co-chair and co-host, and was trying to inject a little unconventionality and risk into what was a very safe and traditional marketing gathering. Knowing my reputation — and the reputation of my book — he brought me aboard, not only as keynote speaker, but also to help in the event's development.

Given the opportunity, I spewed concepts like a punctured fire-hose. Many of these were eventually used, like bringing comedian Howie Mandel aboard to talk about his international "brand," as well as fuelling the usual Q&A sessions with alcohol and calling them "Having a Beer With" (in an attempt to get the industry interviewee to pull the usual stick out of his/her ass and loosen up a bit).

Despite these innovations, the magnum opus — of course — would have to be my keynote. Making it so became an obsessive, un-relenting challenge. The speech's focus was to be the underlying theme of my book (a somewhat new marketing theory, well-researched and with breakthrough potential), but the fixation to turn it into some-thing spectacular and memorable on stage became a venomous snake with two snarling heads named "intimidation" and "reputation."

The intimidation factor was external, unknowingly driven by the others also scheduled to be on stage at the event. In addition to Howie, the two-day agenda was packed with luminaries like media/retail giant Debbie Travis, acclaimed academics like Dr. Ken Wong, busi-ness legends like Terry O'Reilly, and marketing gurus like Google's Avinash Kaushik. Gulp.

The reputation element was entirely internal. My "personal brand" as "A Somewhat Smarter Smart-Ass" (the result of co-founding the Just For Laughs Comedy Festival, I suspect) combined with the "I am not worthy" insecurity from being part of this all-star lineup drove me to push way, way beyond my usual unconventionality. So, despite my initial respect for those I was sharing the stage with, I now needed to regard them with rivalrous contempt. They had become the competition. And with the fanatical need to overshadow them, I could not be shackled by the bland constraints of a seated interview or the tired PPP format (show your PowerPoint, pace the stage, and pull out the occasional prop) of a customary marketing speech. That may be good enough for gurus, stars, academics, and legends ... but not me — Oh, no!

Only one problem: What would my show-stopping, scene-stealing, competition-crushing idea be? No pressure. As is my habit when it comes to idea-generation, I didn't go "looking" for it; I left it up to fate to let it find me instead. And — voilà, as usual — the inspiration soon came, a concept seeded by a pair of legendary pop songs (if you can believe it).

On stage throwing money at Juste pour rire. Never do anything traditional or boring, I say!

The first stimulus hit me on an early fall afternoon while cruising the city in my 1960 Corvette convertible. I have a strange musical conceit with this car; I will only listen to songs from 1958 through 1966 while driving it, and I loaded up an iPod Shuffle with about six hundred tunes from the era for the purpose. One of them, which randomly popped up on this fateful day, was Peter and Gordon's "World Without Love."

"Hmmm," I thought, "What if I tinkered with the chorus: 'I don't care what they say / I won't stay in a world without love?' What would life be like in a World Without Surprise?"

(A little aside when it comes to failure; rarely is it the fault of "one thing." More likely, failure is the end result of a series of events, with its seeds often planted far, far away on both a time and physical scale. When it comes to My Greatest Failure, embracing Peter and Gordon's tune was the ground zero seedling.)

Bang! The first stake sledgehammered into the ground! A world where there is no such thing as surprise! An inspired concept, for sure — but what to do with it? This is when things started to get somewhat wacky.

Given my disdain for the ordinary, and the driving need to make a CMA splash, I turned up my nose at the usual presentation mode of slides and bullet points, and once again let fate — and "World Without Love" — drive my destiny. It seemed only natural that if the inspiring song's lyrics were in the form of rhyming verse, my speech should be as well.

That's when my surprise speech became a surprise poem. It made perfect sense! You can't speak about surprise by rote: "Here's the first point ... here's the second point ..." and so on. No, no — you gotta do something surprising! And what will be more surprising to a gathering of the country's top agencies, marketing execs, and brands than a business presentation in the form of a poem?

Uh — how about a poem with choral support?

No joke. A few days after writing the eighteen-stanza (!!) rhyming, rhythmic World Without Surprise manifesto, while in the shower, I heard the Stevie Wonder tune *Love's in Need of Love Today*. The haunting choir that underlined the song's opening and first few verses dropped my jaw. "That's it!" I said in a "Eureka!" moment. "The soundtrack to the poem!"

Another series of seeds planted. A gospel choir would back me up with a mournful droning through the first twelve dystopian stanzas of the poem (outlining the dreariness of a world without surprise), and then break out into an upbeat hallelujah for the final utopian six (focusing on the positive power of you-know-what). All that was left was finding the gospel choir to back me up! (Surprisingly, that part was fairly easy; finding eight matching robes for them was hard.)

So let's recap: No mere marketing speech, but a protracted poem supported by an eight-member gospel choir.

Crazy, no?

Well, yes — but not enough! Consider the CMA audience — mega-marketers. People perhaps even more jaded by "standard stuff" than I was! What's more, they were *powerful* jaded people, with connections and influence that could perhaps vault my "surprise concept," my book, and me into the upper stratosphere! Reciting a musical poem was merely the first step; to *really* impress and move them, it required a level of packaging equal to the somewhat-twisted content.

Divine Heroic Inspiration

One of my heroes is the ground-breaking comedian Andy Kaufman, a renegade whose reality-blurring performance art endeavours (such as lip-synching the *Mighty Mouse* theme to a kiddie record on *Saturday Night Live*, or taking the entire audience of his Carnegie Hall show out for milk and cookies) endeared him to a late-1970s fan base. They may have spent more time scratching their heads in bewilderment than laughing, but people seemed to love Andy, and they waited with bated breath to see what he would do next.

Channelling my namesake's spirit (he passed away in 1984 at the age of thirty-five from cancer, way too early) the CMA poem packaging plan seemed to assemble itself in his image. I would take to the stage, appear somewhat nervous, stutter and screw up — so much so that I would actually stop abruptly, walk off stage, and ask to start all over again.

My goal was not only to destabilize the crowd in a very Kaufman-esque manner ("Uh, is this real?"), but to set up the entrance of the choir. You see, the first six minutes or so of the speech were relatively customary (that ol' PPP process I outlined earlier); performed to tee up the revelation of the poem. When I ultimately made my way over to the podium for the poem's recital, I would "realize" I had left my binder backstage (yet another destabilizing screw-up), and ask for a stagehand to bring it out. Said stagehand would be a choir member, and my commenting on his purple satin robe would be a cue for the others to join him in choral posture and start harmonizing.

At this point, I wasn't merely planting the seeds for failure; I was fertilizing them and watching little vines sprout. But as they say in those infomercials, "Wait ... there's more!"

Cut to Tuesday, May 25, the night before the fateful presentation. As a member of the organizing committee, I was invited to a private, pre-event dinner, at which I was seated next to Farah Perelmuter, head of the prestigious Speaker's Spotlight Agency. We got to talking about what I was about to do, and while I didn't reveal too many of the punches, she seemed intrigued by the promise of something very different.

I was getting greedy. I had already invited my own speaking agency to my gig (attendance confirmed by its president, Theresa Beenken, and a handful of the firm's other agents), but, turbocharged by my own delirium, and tipsy from the ever-flowing wine, I insisted that Farah be there as well. In my delusional mind, I would set up an agency bidding war for the deluge of speaking opportunities that would no doubt be sparked by my performance. Whether she was just being polite or truly interested, she was in the audience the next day.

Not over yet, folks. As the final nail in a coffin I did not yet realize I was building, I had spent the entire amount of my speaking fee on a camera crew to capture this marketing-world-changing event — not merely for posterity, but as a demo reel to help sell it. I went all out in my instructions to ensure that everything and everyone would be filmed — the false start, the faux nerves, the choir, the enraptured crowd, the delighted and soon-to-be-warring agents. Nobody would ever be able to forget this!

The Date with Destiny

I arrived early at the Bassett that Wednesday morning, ostensibly because that's what organizing committee members must do, but I had other things on my agenda. On the practical side, I had a clandestine, locked-door rehearsal scheduled with the choir. I wanted to walk the stage and plant the binder. I also had to hang — and hide — my stage clothes, which were to be the icing on the cake of my extravaganza. (I say this almost literally, as I wore a hot pink Marc Jacobs jacket, a shade usually seen on kiddie confectionary or as gooey swirls on a birthday cake.)

Once all my checklist items were knocked off, I found a seat in the theatre to take in who was on before me, and who was seated around me. They may have been a who's who of marketing thought and royalty, but to me they were all the same — a swarm of smart folks in greys, blacks, and navy blues who were about to be thunderstruck by the 3D kaleidoscopic tsunami I had prepared.

After a procession of speakers, their perfunctory slides and powerful points (as well as a lunch break where I rehearsed with the choir one last time), it was finally my time.

My date with destiny.

My day of reckoning.

Deep breath, as Mitch Joel introduced me …

Twenty minutes later, it was all over. And amazingly, every-thing worked! Went off without a hitch. As planned, I intentionally stuttered and flummoxed my way through the first ninety seconds, stopped, asked to start again, did so, and went through the paces of a "typical speech" for a few minutes. I then made my way to the podium, called for my "forgotten" binder, greeted the robed "stage-hand" and recited the eighteen stanzas as the choir hit all the right notes. Like clockwork!

That was the good news.

The bad news was that "A World Without Surprise" didn't con-nect with the audience one iota. A total miss. I could tell from the opening seconds that they didn't get it, didn't buy into the concept, and most defiantly didn't want to be taken on a ride of something new. I peered past the lights into the crowd and saw a swamp of be-wilderment, indifference, and contempt. The same faces that were enraptured with previous speakers, while hands took copious notes from their PowerPoints, were now sullen death masks.

So what does one do when faced with these types of faces? I, of course, tried even harder to win them over.

To no avail. Not even close.

More Than Mere Hatred

In the end, they didn't just hate the speech, they *fucking hated it*. Let me explain the subtle difference between the two designations, separated by one necessary F-bomb I apologize for having to drop in this book:

> Hate = "You suck. Let's move on."
>
> Fucking Hate = "You wasted our time so thoroughly, we want to kill you and make sure you never get on another stage again."

Worse yet, they didn't only fucking hate the speech, they fucking hated *me* even more. While I was cavorting on stage, the Twitterverse was strewn with the type of vitriol usually reserved for oppressive dictators or dog beaters. They picked on everything — the way I walked, the way I talked, the way I dressed, even my hair. (I had quite a funky haircut at the time, and one of the Tweets assailed, of all things, my "bad toupee"!) A few weeks later, on a wrap-up conference call, I learned that at 15 percent, I had set a record for lowest-ever approval rating in the history of CMA conferences. A sickly farm animal relieving itself on stage would've scored higher. ("Even Hitler did better, at 18 percent," quipped a sardonic committee member. Gee, thanks.)

The seeds of failure had now sprouted into a sequoia-sized sapling reminiscent of Jack's fairy-tale beanstalk. But there was no climbing it to escape. At this point, I still had to face the attendees in the theatre lobby (worse yet, I had to face them at the party that night; being on the committee, it would be unacceptable not to show). In my troubled mind I pictured them maniacally seeking revenge, an angry mob of clenched fists, pitchforks, and torches, like in those old Frankenstein movies.

It wasn't that bad, but close. Let's just say there was nobody hanging around for autographs. From the moment I left the stage, and continuing for the rest of the conference, I endured the gauntlet of shame, where people would separate like Moses's Red Sea as I walked among them, then turn their heads or look at the floor to avoid making even the most perfunctory eye contact with me. The speaking agents — real and imagined — conveniently had "other appointments" to get to and were long gone. Even the choir, so jubilant at every turn, dispersed stealthily and silently.

But as they *don't* say in those infomercials, "Wait ... it gets worse!"

The only person to greet me was the videographer I hired. As mentioned, I had spent every penny I earned for this gig on him and his crew, and paid up front. He had to deliver something. Sheepishly, he placed the DVD evidence of the event into my hand as if it were a fragile vial of highly contagious infectious disease. As I had predicted — with this digital capture, there was now no way this could ever be forgotten.

I took the cursed DVD home and threw it in the bottom of a deep drawer, where, like the *Titanic*, I hoped it would stay buried for years. Maybe not *truly* forgotten, but deservingly ignored.

Alas, is this where the story of My Greatest Failure ends?

Not really. This is where it begins. Go back to my opening line — "This is the true story of a *successful* speech about the subject of failure."

The DVD sat in the aforementioned drawer for six years, covered up by magazines, files, old iPhones, and other miscellany. I never even dared to even peek at it — until I got a call in February 2016 from two guys, Robert Boulos and Francis Gosselin, who were putting on an event in Montreal called "Failcamp."

One day I'll get over it. Trying to erase persistent CMA speech nightmares before hitting the stage at the ResolveTO conference.

Failcamp's slogan was "Failure Is *Always* an Option." Its mission was to demystify and laugh at failure, and it did so by gathering luminaries from different fields to recount personal stories of disappointments, flops, and catastrophes. When I got the call, my mind immediately snapped back to the CMA debacle, and I figured now — a little more than two thousand days after it was filmed — was an optimum time for the world premiere screening of the "World Without Surprise" DVD.

That is, if I could bring myself to watch it first.

This was no easy task. When I fished that disc from the bottom of a now-impossibly cluttered drawer, my stress was palpable. As I slipped it into my old iMac's DVD slot, I felt my gut tighten and revisited the same feverish flop-sweat that overtook my pores at the John W. H. Bassett Theatre when I gazed into the hostile crowd. My fingers trembled as I pressed *Play*.

Cut to the stage of the Sid Lee agency's Innovation Space for Failcamp on Friday, April 1, 2016. The date was very apropos, as eight of us were about to play the fools by reliving our worst professional nightmares, to the delight of a paying public.

When speech failure lives long enough to become speech success. Flying high at Failcamp watching myself bomb at that fateful CMA conference.

The unique event lived up to its name from the moment I walked into the venue. The organizers had screwed up the scheduling, which resulted in some frantic phone calls to me as I sat in a cab at 4:45 p.m., ostensibly well in advance of my 5:15 p.m. start time. "We're running early!" Boulos panted. "Where are you?" This panic led to a wireless mike flung into my right hand like a relay baton immediately upon my arrival; I literally started my speech while entering the building, crossing the front door threshold yapping, still wearing an early spring coat and scarf, lugging a bulging briefcase.

While I had screened the DVD privately, I still had no idea how it would be received in public. But the hype and the presentation around *this* speech of mine worked. Blow by blow, I went through the story behind "A World Without Surprise" to set up the video, and the audience howled upon seeing my jacket, that crazy haircut, and the so-called "nervousness," and, most notably, on every occasion the choir punctuated a stanza with a melodic "No Surprise!"

This time, when it was all over, "A World Without Surprise" was met with a resounding ovation. It only took six years. Timing is indeed everything, I suppose.

Learning a Lesson or Three

I would like to end this story by saying there's a lesson in all this, but that would be selling it short. There are actually *many* lessons.

First of all, I'm proud that I had the guts to try something new back in 2010. Going against the established norm and breaking new ground ain't easy, and while everyone at the CMA event did better than me, *nobody* took a risk even close to the one I did. That's not insubstantial. I may have crashed and burned, but at least I jumped off the cliff.

Second, and most surprisingly, when I finally pressed *Play* on my iMac and watched the CMA speech, I realized ... Hey — it wasn't all that bad! We always make things worse than they are. Granted, it wasn't the greatest public sermon on record, but the surprise theory made sense, I made some valuable points in explaining it, and the poem was clever. Okay, maybe the choir was over the top, but at least there was meat on the bone of the content. No offence to the CMA audience, but they were way more middle-of-the-road than I ever imagined they could be, and thus weren't prepared for someone like me doing something like that.

The main lesson, though, is that failure is a temporary state. It doesn't last. Any failure, no matter how painful or punishing, just doesn't matter in the end. Never mind "in the end," it doesn't matter in six years — even sooner. Time indeed heals all wounds. There is no such thing as enduring failure; its power and status diminishes over time. Yesterday's failure, the one that seemingly destroyed your psyche and self-confidence, is today's story you'll laugh your ass off at over beers in a bar (or at a Failcamp). It's no longer a source of pain, but of hilarity.

Let me go one step further and say that failure shouldn't be feared, because it's a necessity. In fact — hold tight — it should actually be welcomed. A world without surprise pales against a world without failure; we'd become weary and bored with constant success. Success would lose its value. When you think about it, you realize it's the failures that define life's successes.

I learned this welcome lesson first-hand when I was president of the Just For Laughs Comedy Festival. We lived by a curious credo: *If everything worked, we didn't go far enough.* If we didn't fall on our asses, and hard, at least once every year, we would know we had played it too safe. And playing it safe is not good for a business that needs to grow. So failure is your friend, not your enemy. It just takes a while to appreciate.

That's why I used the adjective "Greatest" instead of "Worst" to describe the failure that marks this story. My speech was an epic, complete, unmitigated disaster; even the smithereens were smashed to smithereens. A loud and legendary misfire, to be sure. But if you're gonna go down, it's best to do so in a most remarkable fashion. Enjoy the ride, even if it's on the *Highway to Hell.*

Easy to say, but not easy to do. Frankly, if I could bend the ears of every Canadian business leader — private or public — and every government leader, these are the words I would fill them with: guaranteed success doesn't make you fearless, taking risks does.

This is especially true in the entertainment business. I recently spoke at a conference at Queen's University and kinda shocked the young assemblage when I said that in ten years there *won't be* a Canadian entertainment industry — because it's all going global. Granted that years of policy and protectionism helped build Canada's cultural sector, but the internet and rapid worldwide digital progress is about to disintegrate it.

In its place, Canadian individuals and groups will be battling it out on the world stage with competition from all over (at time of this writing, Canadian artists like Drake, Justin Bieber, and the Weeknd are doing just that, and dominating the American-based Grammy Awards). You don't

This is a speech from C2MTL. Still took risks, but this one went way better!

win battles by playing it safe, relying on restrictive quotas, or gathering be-hind a collective. You win by swinging for the fences, jumping from roof-tops, and accepting the adrenaline-boosting inevitability of uncertainty.

That said and accepted, what's the legacy of My Greatest Failure? Well, after six years, it was reborn as a successful Failcamp speech, made lots of people laugh and think, and is now the most popular chapter in this book (he says hopefully!). My current speaking agent is even trying to develop My Greatest Failure as a motivational keynote for other conferences, corporations, and schools.

Where else can My Greatest Failure go? Don't know for sure, but I suspect its lifespan is far from over.

The world, you see, is filled with plenty of — wait for it — surprises!

o o o

For more context, I've appended this chapter with the poem in its entirety. Enjoy!

A WORLD WITHOUT SURPRISE

In a world without Surprise
No more "Can't believe my eyes!"
What you'd see is truly — only — what you'd get
Life would plod along as planned
All supply and no demand
Muffled colours, dreary skies
In a world with no Surprise

In a world without Surprise
No more sevens or snake eyes
All casinos would go bankrupt in a day
Every bet would be a sure one

Games of chance indeed a poor one
Vanished long shots, hushed loud cries
When the world has lost Surprise

When the world knows no Surprise
Is when competition dies
Our sporting life would go on life support
Games would lose their sense of fun
Before played, we'd know who won
Empty seats, no pennant flies
In the world of no Surprise

What is life without Surprise?
Boredom takes the Nobel Prize
Crackerjack is simply popcorn and glazed nuts
Nothing hidden deep inside
Curiosity denied
Revelations stigmatized
In a world of no Surprise

In a world with no Surprise
Our rights we would compromise
Political campaigns would be extinct
Who would win? Foregone conclusion
Democracy just an illusion
Banana Republics arise!
When the world deserts Surprise

A new world without Surprising
Would see an internet revising
Without shock there wouldn't be Web 2.0
What spreads? Not the mundane
But the wild and the profane

YouTube clips won't tantalize
A new world with no Surprise

No Surprise, it also means
Death of screenplays on our screens
We would know how every movie meets it end
Cliff-hangers searching for a cliff
Stop your wondering "What if?"
Total Hollywood demise
In a world without Surprise

In a world where shock is muted
Richard Branson: three-piece suited
All Seth Godin's cows would be black, brown, or white
We'd hear whispers from Tom Peters
"Made to Stick"? The concept teeters
All this wisdom now unwise
In a world with no Surprise

When Surprise has turned to vapour
Books would be heaps of scrap paper
No more "Catch" in Joseph Heller's 22
What's the use in bookstore spending
When we know each volume's ending
Pens and keyboards paralyze
In a world without Surprise

When Surprise is just a rumour
We would lose our sense of humour
Every chicken too afraid to cross the road
No more punch lines; only set-ups
No more gasp-inducing get-ups
Tears of joy sucked from our eyes

In a world without Surprise

When Surprise does not exist
Every mystery is missed
It's clear just "who-dunwhat" in each "who-dunnit"
All transparent, none opaque
No one hiding in a cake
The unknown exposed by spies
In a world without Surprise

To eliminate Surprise
Is to minimize life's highs
Why must we know tomorrow yesterday?
Our phones tell us who is calling
To not know seems lame and galling
We can't hide, there's no disguise
In a world with no Surprise

BUT THANKFULLY ...

It's Surprise that drives our dreams
By expanding life's extremes
A perpetual discovery of new
By embracing unexpected
You'll live life turbo-injected
Always more to publicize
In a world rife with Surprise

With Surprise firmly engrained
Our emotions, unrestrained
Generate the tales that never fade away
Anecdotes that flabbergast
Memories that ever-last

People upbeat and enthused
In a world Surprise-infused

Far from frivolous, Surprise
It's the Pow! between our eyes
It's the glue that bonds two parties into one
So effective, near perfection
At establishing connection
Bonds with clients crystallize
In a world filled with Surprise

Such importance, this Surprise
Brings us wonder, kid-ifies
It brings every day a taste of Disneyland
Makes your eyes expand and pop
Takes your jaw and makes it drop
Fills your gut with butterflies
That's the power of Surprise

Surprise works, defeats resistance
Gives emotions great assistance
Makes it easier to get a message through
It increases happiness
It's the "Lubricant to Yes"
Turns "Just looking" into buys
Loads of profit in Surprise

As this manifesto ends
I must ask you this, my friends
Are you wondering how it will come to close?
No cheap shock like "Made ya look!"
No lame shilling for my book
In fact, no last line at all ...

6

DR. FRANK PLUMMER

Distinguished Professor, University of Manitoba

Failures in Public Health Science ... and When Failures Lead to Success

I met Frank through his amazing wife, Jo Kennelly. They reside in Toronto, have a great family, and are part of the "proud Canadian" club — even though Jo is a Kiwi. (But we won't hold that against her!) Frank — well, where to start? He is a 2016 recipient of the Gairdner Award, he led the Winnipeg-based National Microbiology Laboratory for fifteen years, he is a 2014 winner of the Killam Prize — the list goes on, and on, and on. I was thrilled when Frank agreed to work with our team and write a chapter with his thoughts on failure in our national health system, along with personal anecdotes from a long and famed career.

— Alex Benay

Failure is often a normal part of health science research. It can lead to success, as was the case with the development of Canada's Ebola vaccine (VSV-Ebov). At the other end of the spectrum, failure can lead to widespread harm and loss of life. There are failures that are the result of deliberate action, others that are the result of accidental action, and yet others that result from the absence of preparedness and response. Individuals, institutions, and systemic issues may all separately or collectively contribute to a failure in the health sciences.

Five examples are explored in this chapter. The first three relate to my work as head of the National Microbiology Laboratory (NML), Winnipeg, and illustrate how success was achieved out of failures. The fourth example illustrates a case of failure, from deliberate inaction, to take account of existing science. The final example illustrates what can happen in the absence of science when the public health system is under pressure, as is the case during a pandemic.

The Role of Federal Politics: CF-18 Fighter Jets and the Winnipeg Lab

Canada is a vast country of provinces and territories, often with competing interests. Rivalries between regions play out on the political stage in Ottawa. The decision to locate the National Microbiology Laboratory in Winnipeg, Manitoba, far from its overseer, Health Canada, arose out a failure to keep a political promise in 1986 — to award the CF-18 fighter jet's twenty-year, $1.4 billion maintenance contract on merit and through a competitive process. The political decision by the Brian Mulroney government to award the maintenance for Canada's new CF-18 to a Montreal firm over Winnipeg's Bristol Aerospace Ltd. was a devastating blow to Manitoba's aerospace industry and economic prospects.

This decision to award the CF-18 contract to Montreal caused consternation and outrage in Manitoba, from Premier Howard Pawley down. It led to animosity between western Canada and Ottawa, impacting the Meech Lake Accord and contributing to the dramatic loss of seats by the Progressive Conservatives during the 1993 national election. It was an outrage few Manitobans have forgotten and has become part of the makeup of Manitobans. Some say that to understand Manitobans you need to understand and know the CF-18 story.

Jake Epp, the health minister and senior Manitoban minister in the Mulroney government at the time, was in the hot seat. As shocked and surprised by the decision as all Manitobans were, Jake Epp knew he had to deliver something for the region. His own department, Health Canada, together with Agriculture Canada was working on proposals to replace their existing Ottawa labs and increase the biosecurity level of pathogens that could be studied and tested in the new labs. Both departments saw the need to develop containment level 4 labs, a level of biosecurity that would be a first for Canada. Given that the departments were headquartered in Ottawa, the capital city was the logical location for a new lab to serve Canada — up until the CF-18 decision.

Epp, after being approached by Manitoba health-research heavy hitters — Dr. Henry Friesen, then head of the Medical Research Council, and Dr. Allan Ronald, world-renowned infectious disease expert — sought a commitment from the prime minister and his Cabinet colleagues that the new, yet-to-be-built Canadian Sciences Centre for Human and Animal Health (within which the National Microbiology Laboratory is housed) be located in Winnipeg, as a quid pro quo for the city losing the CF-18 maintenance contract to Montreal. This effort was supported by Mayor Bill Norrie, newly elected Premier Gary Filmon, and the Winnipeg science and business community, including Dr. Arnold Naimark, the Richardson family, Sandy Riley, and Don Leitch. This centre was to be the first facility in the world that housed both level 4 animal and level 4 human high-containment labs, and remains the only facility of its kind globally. At a cost of $189 million in 1999 ($276 million in 2017 dollars), it was one of the largest investments the federal government had made in Manitoba.

Yet, despite the obvious political climate at the time, bureaucratic opposition in Ottawa to locating Canada's main microbiology lab and new level 4 facility in Winnipeg was fierce. Arguments advanced against Winnipeg included recruitment challenges; size of the city and depth of the talent pool; fear of opposition from the community; it was out of Ottawa's span of control and reporting; and the federal government would experience a brain drain of expertise.

As it turned out, only around twenty-five Ottawa employees accepted the offer to move to Winnipeg. Many lab positions went unfilled. The lab was chronically underfunded for its operational mandate. Many predicted the lab would become a white elephant on the wind-swept prairies. It would follow in the footsteps of Montreal's Mirabel Airport project, which had opened in 1975 at a cost of more than $2 billion (2017 dollars). Mirabel was meant to serve passenger flights but never really did. Today it is used for cargo, and it is widely considered a government funding allocation error.

The Canadian Science Centre for Human and Animal Health in Winnipeg. It houses the National Microbiology Laboratory.

So, How Was the White Elephant Avoided?

The NML's ultimate success was assured by timing, leadership, and location. The timing was right for the expansion of infectious disease effort and funding in Canada, as new biological and bio-terrorism threats arose, penetrating the political consciousness of the nation. Just pre- and post-9/11, Ebola, Marburg virus, and smallpox level 4 pathogens were high on the list of potential biological weapons of terror. Canada — without a level 4 lab of its own — was dependent on the United States for its knowledge, intelligence, and development of effective field and therapeutic responses. None of which would have been a viable political and national response in the event of an outbreak in Canada of level 4 pathogens or a bio-terrorism attack.

With political and scientific interest in high-containment research and the move to Winnipeg settled, the new deputy minister David Dodge — a gruff, visionary leader — resolved the lab's funding shortfalls and established a blue-ribbon panel to recruit a globally recognized scientist to build and lead an "outstanding scientific program." The blue-ribbon panel advertised widely, took several months to find the right candidate, and worked with the University of Manitoba to ensure that salary was not a barrier to getting the right talent for the task. This process resulted in my appointment. At the time, I was working in Nairobi, Kenya, running the world-renowned Kenya AIDS Control Program, and as a consultant to the World Bank, World Health Organization, and numerous governments in developing nations. As a Winnipeg-born scientist and product of the University of Manitoba, my appointment was welcomed with pride and excitement by the community. It was front-page news in Winnipeg.

The NML became a strong magnet for doctors and scientists from elsewhere in Canada and from around the world, including the United Kingdom, the United States, and Germany. A key early recruit was Dr. Heinz Feldmann, M.D., Ph.D., the now-famed inventor, along with Steven Jones and Ute Ströher, of Canada's Ebola vaccine VSV-Ebov. Feldmann was there ten months before my recruitment. He was brought onboard to build the level 4 program and prepare for this lab to "go hot" — work with some of the world's deadliest pathogens, such as Ebola, Marburg, and Lassa fever. Feldmann built a stellar and globally influential program. Eventually, research fellow Dr. Gary Kobinger took over leadership of the program. Gary continued with the same high standard of scientific excellence that Heinz demonstrated, including developing ZMapp — a cocktail of three monoclonal antibodies given to patients post-exposure to Ebola — with international colleagues. Sustaining funding for the science was a continued struggle. Described by Health Minister Tony Clement as "the jewel in the federal health portfolio's crown," the NML has been fortunate to receive the confidence of the city, province, and federal government. Through five prime ministers (Brian Mulroney, Jean Chrétien, Paul Martin, Stephen Harper, and Justin Trudeau) and fifteen health ministers, the NML has grown into a centre of scientific excellence and international influence.

As they say in real estate — location matters. The support and approval from Winnipeg City Council to locate the lab downtown, in a residential community, only a few hundred metres from the Health Sciences Centre, Canadian Blood Services headquarters, and University of Manitoba Medical School, enabled a close exchange of ideas and collaborations directly with the clinical teaching community. It supported the co-location of research and adjunct appointments with the University of Manitoba, a known powerhouse in infectious disease circles. Together, this cluster created a culture of scientific excellence and global responsibility. The best researchers in the world knocked on our doors seeking postgraduate opportunities and employment. There's a pride in working at the NML, and the people we employ want to stay because of the level of responsibility they can achieve and the flexibility of the scientific program. It costs millions of dollars to train a level 4 scientist, and graduates are required to work long hours and in dangerous places around the world. Federal salary scales, below provincial scales and salaries in the United States and Europe, are why many good scientists choose to leave the service of the federal government.

"Shipping Issues" Are a Lame Excuse ... When the FBI Calls

It wasn't long before the location of the lab in Winnipeg delivered its first failure, necessitating a decision on my part to act quickly. With the decision having been made to locate the lab way outside of the Toronto-Montreal-Ottawa corridor, and some distance from the smaller Atlantic provinces, and with the pride of Manitoba resting my shoulders, it was my job to prove the critics wrong.

"This is the FBI," came the voice on my phone one Wednesday morning just a year after starting the job. My ever-present flip-phone lit up as I entered the NML via the rear carpark entry. A man with an American accent confirmed that I was "Dr. Plummer," provided his credentials, and then let me know that the FBI had become aware, via back channels, that a suspicious white powder had been found in an envelope in a hospital in New Brunswick.

Alarming enough for those who received the envelope and opened it, the white substance evoked heightened concern for two reasons. The envelope had arrived in the mail from Chile — the source country of anthrax-laced mail that had been sent to the United States shortly after 9/11 — and it had been opened in a space that shared its heating and air conditioning system with the hospital's intensive care unit and oncology ward. Exposure to patients in these wards, already in a weakened state, would certainly result in the loss of life if the substance entered the hospital's ventilation system.

Once we confirmed the call, we contacted the public health authorities in New Brunswick. They went to work to courier a specimen of the powder to the NML, a trip that would take a few hours with Winnipeg more than two thousand kilometres away. Time passed quickly that day. By the end of the business day, the specimen had not arrived. Nor did it arrive by the end of the day on Friday — sixty hours later. The weekend was looming, and the specimen had not yet reached Winnipeg. "Shipping issues" was the reason given to me for the delay.

With political pressure for answers mounting, "shipping issues" was a pretty lame response, and not something I was willing to communicate to Ottawa regarding what was potentially a national crisis, and before I had taken action to correct the situation. Managing panic at higher levels is one of the key jobs of the head of lab during a public health crisis.

With my head on the political chopping block, and the pride of the province resting squarely on my shoulders, the adage "it is easier to ask forgiveness than to beg for permission" guided what I did next. In fact, it became a necessary part of the job. With lives on the line, the "Oh shit!" moments of my job require that I gather the facts as quickly as possible and use my judgment to direct the necessary action as soon as "the call" arrives at the lab. Lab staff jumping on planes only to be called back later has to be built into the response capabilities of the lab, especially when a verifiable threat exists to people's lives. Action calms heads, too.

So, Saturday morning I woke early, called lab staff at home, asked them to arrange babysitters if necessary, and chartered a jet plane from the phone book's Yellow Pages. In just three hours, a quickly assembled three-person team, led by Dr. Amin Kabani, with an equally rapidly assembled portable testing lab, was on its way to New Brunswick. I must have sounded trustworthy on the phone, because the jet charter company agreed to my request based only on my word: no purchase order, no cash deposit, and no credit card guarantee — just Frank phoning to charter a jet plane for the weekend.

Thankfully, testing proved the New Brunswick hospital powder to be safe — not a threat to human health.

Rapid Response Teams and Mobile Labs (Lab in a Suitcase, in a Shipping Container, and on a Truck)

After this episode — this expensive and frustrating failure to quickly get our hands on a potentially deadly substance and test for radioactive, chemical, and biological agents — I decided, along with Dr. Neil Simonsen, Dr. Amin Kabani, and others, that it was necessary for the NML to be ready on three hours' notice to fly anywhere in Canada or around the world. If called upon, Canada would have an infectious disease rapid response team, able to respond to disease outbreaks and bio-terroism events from the Arctic to the Antarctic and

The NML mobile laboratory setup.

everywhere in between. We had the capacity, skills, and resources to halt an outbreak at source in poverty-stricken regions of the globe. I decided it was my mandate to do so. The most effective way to protect the health of Canadians is ensuring deadly organisms never land here.

Today, the NML has two teams of three responders on permanent standby, with their equipment and supplies stationed at the Winnipeg air force base. Our "lab-in-a-suitcase" capability was designed to fit into thirteen suitcases that could be checked-in on a commercial flight and easily transferred to a much smaller airplane that could fly and land in remote locations. For years Canada was the only country with this capability. The model has since been copied by other countries, and it has transformed the way the World Health Organization (WHO) and Médecins Sans Frontières (MSF) respond in the field. The team, led by Dr. Heinz Feldmann, and later Dr. Steven Jones and Dr. Gary Kobinger, has made numerous missions on Canada's behalf to provide onsite testing for viral hemorrhagic fever and infectious agents, including travelling to Sierra Leone (Ebola), Democratic Republic of the Congo (Ebola), Angola (Marburg), Iran (Crimean Congo Hemorrhagic Fever), and Bangladesh (Nipah). A team was deployed in 2003 to Hong Kong to support the World Health Organization and Hong Kong government with the investigation into the Amoy Gardens and Metropole Hotel SARS Co-V outbreaks. These were the origins of the Canadian and global SARS outbreaks.

During the 2009 H1N1 outbreak, specimens for influenza testing arrived from Mexico on President Calderon's jet. The call from the Mexican president's office came in on my phone after the plane had taken off. I quickly had to arrange with Minister of Health Leona Aglukkaq, via her chief of staff Dani Shaw, for a call to be made to Minister of Defence Peter MacKay's office to receive clearance for the plane to arrive at Winnipeg's military base after midnight. The plane arrived at 1:00 a.m., and we met it with two teams — one to off-load and secure the specimens to take to the NML for immediate testing, and the second team to fly to Mexico on the return flight to assist with the outbreak control effort. Led by Dr. Ute Ströher, this team set up diagnostic tests and a public health intelligence database to monitor and track patient numbers and disease mutations in real time.

Furthermore, a $3 million operations centre was built on the main floor of the building, lit with natural light (and near the cafeteria), and designed to be switched on at a moment's notice. Its purpose is to coordinate communications with provincial labs and provide support to staff in the field — everything from shipping specimens, arranging travel, and coordinating data, results, and progress related to the activity at play. Over half of the staff that work at the NML are trained to work in the operations centre, with a roster of names and schedules. A red phone connects Winnipeg with Ottawa.

A high-containment lab on the back of a Mercedes Benz truck base was built to provide onsite testing under special circumstances.

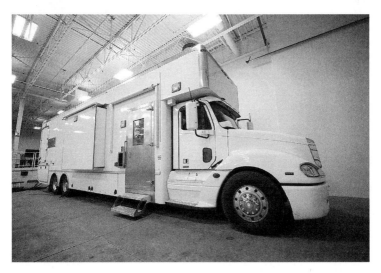

The NML's lab on a truck.

Dr. Cindy Corbett and her BADD team in the lab on a truck.

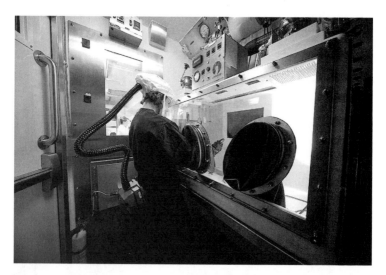

Working in the high containment portion of the lab on a truck.

Used in planned deployments, such as the 2010 Olympics in Vancouver or the G8-G20 in Ontario, the truck is a fully functional level 4 lab on wheels. The design and build lead was level 4 scientist Dr. Steven Jones.

It can travel anywhere in the Americas, although reasonable roads are required for it be fully deployable, and driving it takes time. For more-difficult-to-get-to areas and faster deployment, a lab in a shipping container was designed and built that could test level 4 pathogens and could be flown, in theory, on a C130 or C17, towed behind a transport vehicle, or loaded on a flatbed using commercial forklifts. A team to respond to bacterial antigens was also established. Led by Dr. Cindy Corbett, it was called the Bacterial Antigen Diagnostic Directorate, or BADD.

In spite of several obstacles at the start, decisive action and a creative team of scientists and inventors at the NML turned failure into a success pretty much right out of the blocks. The Winnipeg location, far from political and bureaucratic masters, turned out to be an advantage, suitable to my management style of embracing a larger purpose and taking considered action based on medical training and public health experience. The location of the lab at mid-point of the country now supports rapid deployment nationally and globally.

The Canadian Ebola Vaccine — VSV-Ebov

Ebola is a hemorrhagic fever virus. It looks like a thread under an electron microscope. Ebola was discovered in 1976 in a remote region of the Democratic Republic of the Congo. It is transmitted by close contact between people, through exposure to blood and bodily fluids, including contact with recently deceased persons. Ebola causes systemic infection that results in generalized hemorrhaging — bleeding from the eyes, nose, and mouth. Up to 90 percent of people with Ebola die. It has never been seen in Canada, but it could arrive here any time.

Circa 1999, Dr. Heinz Feldmann set out to study the pathogenic effects of the Ebola glycoprotein (Ebov). He believed this molecule was the cause of the severity of Ebola infection. Early experiments to test this hypothesis failed. When mice were inoculated with the Ebola glycoprotein, using a viral vector, Feldmann and his team discovered that when challenged with the Ebola virus the mice remained healthy, instead of developing Ebola as they had expected.

Upon further investigation, what at first had seemed like a failure turned out to be one of the most significant breakthroughs in Ebola vaccine research. *Fifteen years before the 2014–15 Ebola outbreak in West Africa*, Feldmann's team at the NML had unexpectedly discovered a potential method for Ebola immunization and protection; albeit, some of the early work was also intended to protect the great apes that were at greater risk of Ebola than humans, and were also a potential pathway to negatively impacting human health.

In 2000, with these research results in hand, Heinz approached me with the idea of developing an Ebola vaccine. I thought it an exciting scientific endeavour and decided to get behind it. Anyone who has spent time in Africa understands the global health implications of a virulent agent like Ebola exploding from small community outbreaks to major epidemics. Watching HIV for two decades, I believed it was only a matter of time before an Ebola outbreak happened. Added to this, Ebola was widely regarded as a bio-terrorism threat and, indeed, it had been weaponized by the former Soviet Union.

However, there is not an unbroken, straight line connecting an immune response in mice in Winnipeg, Canada, to a safe, injectable Ebola vaccine for humans in West Africa. Many systemic failures stood in the way and slowed progress.

First, while basic operational resources existed for the level 4 program, large-scale funding necessary to undertake a vaccine program was not available. To proceed, we cobbled together funding over many years from various granting agencies in Canada and the United States, including Defence Research Development Canada (DRDC), the Canadian Institutes of Health Research (CIHR), the U.S. National Institutes of Health (NIH), and internal NML sources.

We first presented the data that showed the vaccine worked in 2002, and then moved quickly to attempt to secure funding from the DRDC and other sources. However, despite being the highest ranking DRDC CRTI project that year, the vaccine project was not funded. ("CRTI" is the stripped-down acronym for "Chemical, Biological, Radiological-Nuclear, and Explosives Research and Technology Initiative.") A decision was made that the overall costs of developing a vaccine from lab to clinical practice would mean it would be never be put into use. It took several years to secure project funds, and then more time to put contracts in place and secure sign-off from Ottawa. DRDC's rationale for turning the project down at the time was an understandable one — especially as market failure was an ever-present part of the Ebola vaccine story, as we shall see. If it had not been for the persistence of Steven Jones, Ute Ströher, Dorothea Blandford, and myself, the vaccine program would never have got off the ground. Steve and Ute, hot off an outbreak in Angola in 2005 and frustrated by the lives lost, did not let up on the importance of the program. Dorothea did yeoman's work, aided by Brian Szklarczuk, my assistant Melinda Patterson, and others, going back and forth with Ottawa and private-sector companies on contracts related to the vaccine licence and development of clinical trial lots.

Second, and related to vaccine project funding, inquiring minds at the bureaucratic level in Ottawa routinely questioned the value of the lab spending its limited resources on Ebola research, when more pressing public health issues and concerns in Canada were seen to be of greater importance. Rather than getting behind this vitally important vaccine research, departmental staff often challenged it during budget discussions; indeed, the value of the entire level 4 program was often questioned with the churn of assistant deputy ministers through the department. I was convinced of the value and pushed for its continued support. However, successive prime ministers and ministers of health who visited the NML thought the Ebola work was pretty awesome! Throughout the debate on restrictions on federal scientists, I often remarked that we at the NML felt nuzzled, not muzzled, by politicians in Ottawa.

The third failure was the anticipated market failure of vaccine development for diseases affecting a small group and occurring in remote regions of the world. Prior to the 2014–15 Ebola outbreak, there were fewer than four thousand cases recorded since its discovery. Many requests for support from Canadian biotech companies and international pharmaceutical corporations were met with a lack of interest, despite exceptionally encouraging results. After lengthy and extensive search, one small U.S. biotech company, Biologic Protection Systems (BPS), showed interest in licensing the vaccine, now called VSV-Ebov. The deal structured was that we would fund and develop the clinical trial lots, and BPS would raise the capital to fund the clinical trials.

Even with a licence agreement in hand, finding funding to support the development of the clinical trial lots was a struggle. The primary inventors of the vaccine had left the NML, taking up leadership positions elsewhere in Canada and the United States. I became the "Scientific Champion" (overall scientific leader) for the Ebola vaccine project, and again cobbled together funding from various sources. A further market failure we discovered was that the capacity to make the clinical trial lots was severely limited in North America. As a result, we looked to Europe — contracting several companies in Europe to manufacture the component parts required for the human-grade vaccine. The process of deal-structuring and licensing agreements, identifying manufacturers, and production took almost three years, and was superbly managed and coordinated by Dr. Judy Alimonti. Judy scoured the earth looking for companies to manufacture the component parts of the vaccine, and then once the trial lots were developed, tested them in animals, setting the stage for field use of the vaccine.

Fortuitously, given the above systemic failures, human-grade clinical trial lots were ready early in the outbreak and were part of

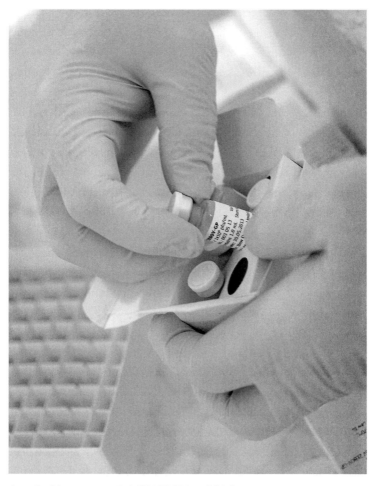

A vial of human grade VSV-EBOV in 2014.

Canada's significant contribution to the global response to Ebola infections. One thousand doses of the vaccine were donated to the World Health Organization. Initially, they were stored in Winnipeg before being moved to Ottawa, and before transfer to the WHO.

The failure to move forward more quickly with clinical development of the vaccine and the licensing agreement later attracted criticism from a small group of academics, unfamiliar with the details of the science and licensing agreements, the scientific mandate of the NML, and the market failures and obstacles the project had successfully tackled. The lack of a usable vaccine on a large scale prior to the 2014–15 outbreak in West Africa cost lives. And, though there were many "failures" along the way, no other government in the world has achieved what Canadian scientists did — clinical trial lots ready for donation at the start of a global pandemic. Subsequently, 100 percent efficacy of this vaccine has been proven during the expedited ring vaccine trial in Guinea supported by the global community. Imagine if the world had had clinical trial lots ready at the start of the HIV pandemic. How different the world would be today!

Since the initial West African outbreak in 2014, VSV-Ebov has been used to combat a small Ebola outbreak in Guinea, in March 2016. The "Canadian vaccine," as it is known, was administered by health workers to patients. Royalties paid to the federal government for licensing the vaccine has thus far exceeded several million dollars. The importance of work on infections like Ebola has been acknowledged more broadly. The Bill & Melinda Gates Foundation and the global development community announced in Davos this year the establishment of a $1 billion fund to create new vaccines for emerging infectious diseases, targeting pathogens for which few or no medical countermeasures exist to combat them. The world has learned from the Ebola experience, with Canada showing the world what is possible scientifically. Small investments in the Ebola program over successive governments produced two novel and effective Ebola countermeasures (VSV-Ebov and ZMapp), and a rapid response team second to none globally, from a handful of intensely dedicated scientists.

Canada's vaccine story is a success made by overcoming systemic failures and obstacles. It is something of which Canadians should be very proud. It is a testament to the excellence of science being pursued by federal employees, and to the lab's location in Winnipeg. Far away from curious eyes, I was able to set the scientists free to do exceptional work. No one gave us permission to pursue Ebola vaccine work. We just did it.

Blood Crises — A Monumental Canadian Failure in Adjusting Practice to New Science

One of the greatest preventable tragedies in the history of Canadian public health is the blood crisis of the 1970s to the 1990s. The Canadian Red Cross, responsible for blood collection and distribution since the 1940s, failed to adopt screening for hepatitis C and HIV. This led to thousands of Canadians contracting these diseases through tainted blood transfusions, despite routine diagnostic tests being available for both. In fact, during the four-year period the federal government argued the agency had responsibility (1986–90), more than 1,100 Canadians contracted HIV, and over 20,000 were infected with hepatitis C, all from transfusions of tainted blood supplied by the Red Cross. An especially notable failure was that blood taken from prisoners infected with hepatitis B was distributed from the 1970s though the 1990s.

As a result of these avoidable lapses by the Canadian Red Cross in screening donated blood and the large number of Canadians affected, the Canadian government established a Royal Commission in 1993 to investigate how the system of government, private, and non-governmental organizations responsible for supplying blood and blood products to the health care system had allowed contaminated blood to be used.

This commission was headed by Justice Horace Krever, with the final report of the commission tabled in the House of Commons on November 26, 1997, after more than four years of testimony, study, and consideration. The *Krever Report* found negligence at multiple levels of the Canadian blood system. Several recommendations were made, including the creation of the Canadian Blood Services for all provinces and territories, except Quebec, and Héma-Québec for Quebec. Thirty-two criminal charges were laid. Compensation for those affected (survivors and families) came much later. People infected with HIV, or their families, were compensated in 1991 (many had died, and most were dying). Hepatitis C victims has to wait until 1999 for those infected between 1986 and 1991, and 2006 for those infected before 1986 and after 1991.

The overriding failure was the unwillingness of the Red Cross to incorporate science into its screening and supply processes. Screening and storage practices lagged behind the development of reliable diagnostic tests for hepatitis B, hepatitis C, and HIV. The core antigen hepatitis B test was available in the late 1970s, but despite being a reportable agent under the Food and Drug Act, testing was not implemented in Canada. In 1984 researchers in the United States had discovered that heat-treating blood killed hepatitis C, HIV, and other pathogens. The Red Cross was slow to implement this, and when it did heat-treat, it didn't treat all blood products. It failed to treat the tainted stock it had in hand, citing budget reasons. Heat-treating the existing stock would have cost $2 million.

Furthermore, screening practices lagged behind epidemiologic data. It was well known from epidemiological studies that hepatitis C and HIV could be transmitted via blood products, with elevated risk of transmission among certain groups, namely homosexual men, injecting drug users, and Haitians. Despite this, the Red Cross stated that "there is currently no available evidence that blood transfusion recipients are at a higher risk of developing AIDS." A fear of exacerbating blood supply shortages and accusations of racism from the Haitian community in Montreal was partly behind the decision to ignore the science.

The same failings can be said of the regulator — the federal government. Health Canada, as Krever noted, should have been ensuring the Red Cross took existing science into account. Further, a disconnected surveillance system across the country, together with varying responses by the provinces, made identifying infected individuals more difficult than it needed to be.

Since this period and the establishment of the NML on the same plot of land as Canadian Blood Services (CBS), the federal government works with the Red Cross through the NML to ensure it is aware of the latest infectious disease threats to the blood supply. The NML, for example, was instrumental in helping CBS establish diagnostic practices for West Nile virus and Zika. The NML is a strong voice for infectious disease control that was not there when the blood crisis arose.

Clinical Trial Failures

Failures in clinical trials are commonplace. A study in *Nature Biotech* found that only one in ten drugs make it to market. Only 32 percent of drugs have the chance to make it to phase III clinical trials. The cost of clinical trials research is high, and the cost of failure impacts research programs, reputations, and patients. Pandemics bring a new type of failure with them. Under pressure from political and government bodies to find a solution, good science gives way to sudden support for hunches. Findings are broadcast by press release, rather than through peer-review processes. Two notable failures of this type included the clinical trials of cyclosporin for HIV/AIDS patients 1985–86, and the use of interferon during the SARS and H1N1 outbreaks.

HIV/AIDS — The Use of Cyclosporine for Halting the Progress of AIDS

The 1980s were a challenging time worldwide for certain segments of the population. A disease that came to be known as AIDS started striking sexually active male homosexuals in North America and Europe, hemophiliacs, injecting drug users, and Haitians. This threat — acquired immune deficiency syndrome — that cost the lives of millions of people worldwide was found to be caused by HIV, or the human immunodeficiency virus. HIV attacks and finally destroys a person's immune system. With a weakened immune system, the patient's body is unable to fight infections and cancer; a host of opportunists eventually invade the body, including tuberculosis, Pneumocystis pneumonia, Kaposi's sarcoma, and lymphomas, to name a few.

Researchers reasoned that a drug — cyclosporine — that had been used for years to prevent the immune system from rejecting transplanted organs would be effective for patients infected with HIV by blocking the viruses' attack on the immune system. Small clinical trials of cyclosporine therapy were conducted in Canada and France in 1985–86.

The Canadian trial took place in Toronto. Nine AIDS patients volunteered and were treated with cyclosporine as outpatients at the Toronto General Hospital. One patient was withdrawn from the study after only a few days because his health deteriorated during the treatment. The remaining eight patients received treatment for an average of fifty-four days each. The Canadian doctors concluded that cyclosporine treatment in patients with AIDS was not beneficial, and in fact that the "drug was demonstrably toxic in a high proportion of the cases." Bluntly put, they concluded that the condition of most of the patients "deteriorated during cyclosporine therapy." The study was stopped. The cyclosporine failure passed into the annals of public health medical research with neither premature fanfare nor political intervention.

Across the ocean in France, a clinical trial of cyclosporine was conducted under very different circumstances. Six AIDS patients were recruited for the study, at varying stages of disease. Less than nine days after the study began, and contrary to the scientific convention of announcing scientific discoveries after peer review by other researchers, the French doctors held a news conference announcing that preliminary tests involving the two patients indicated that the treatment appeared to prevent the virus from multiplying. It had likely proven effective in halting the progress of AIDS. Infectious disease and AIDS experts around the world reacted angrily to this announcement, declaring the results premature. Six months later the same doctors stepped up to the microphone to acknowledge that the cyclosporine treatments of AIDS had failed. The doctors blamed the early announcement on political pressure from the French health minister. Much damage was done to the patients and families involved in this study, the reputation of the doctors, and public trust in the French institutions that housed the research.

While both Canadian and French trials failed to show any benefit in the patients concerned, the French study failed to follow even the most basic of scientific practices. As Dr. Anthony S. Fauci, director of the National Institute of Allergy and Infectious Diseases in Bethesda, Maryland, said at the time, "There's not a scientist I know who'd give something for one week to six patients and make an announcement in the press," adding, "if you want to talk about ethics, you want to make sure something works before you announce it." The failures of the French study went beyond the drug itself. They extended to an absence of accepted scientific practice, reputational damage to the institutions involved, and an undermining of global public health efforts.

SARS and HINI — The Use of Interferons

Another "pandemic failure" involved the use of interferon to first treat SARS in 2003, and then H1N1 in 2009. Interferon medication is a synthetic version of a naturally occurring protein that is involved in the immune system. The body produces interferons to help fight against disease and infection. These proteins stimulate immune cells to destroy cells that have become infected with viruses.

Researchers believed that because interferons had been successful in combating several kinds of diseases, hepatitis C, and certain cancers, it held promise against SARS. A clinical trial to investigate the therapeutic potential of interferon in SARS patients was initiated in Toronto. Doctors were trying all kinds of treatments in an effort to save patients' lives. The interferon study was another stab in the dark. It was approved by Health Canada based on results from a study the year before involving a small sample size and an unusual study design. Significant resources were committed to support this trial. The hope of the interferon response was spread by press release. Scientific shortcuts were taken.

But, it didn't work. The early results were not statistically relevant. Ironically, the study illustrated more than anything else the difficulties of attempting to conduct a randomized clinical trial during an outbreak, especially without the necessary animal and human data to back up the allocation of resources, and when the path of the epidemic was unknown. The SARS outbreak in Toronto was over before robust and valid results were collected. A similar situation arose during the 2009 H1N1 outbreak. An interferon clinical trial was established on the basis of limited science. Not surprisingly, this too met with similar disappointment.

The difference with the development of Canada's Ebola vaccine based on fifteen years of scientific study, an incredible amount of work, and a well-designed clinical trial involving Norway, Canada, and the WHO is stark. It had a huge impact.

Serendipity almost never occurs in the clinical environment. A new approach is not a bona fide innovation until the invention adds value to patient care and health system results by virtue of population-based study. Even during an outbreak it is incumbent upon the regulator to ensure that some scientific evidence is there before patients are subjected to therapeutics based on little more than a hunch. Following these hunches is a failed strategy that diverts resources from other potentially more robust approaches.

Conclusion

Failure is frequent in health science research. It is part of the trial-and-error process of scientific progress. Overnight success is rare. Real success is more likely to be achieved by hard work and dogged pursuit of results, following well-established scientific practices and procedures.

The three examples from the NML in this chapter illustrate how failures can — under certain conditions — lead to success, as well as the importance of hiring the best people and giving them the tools and freedom to do their job, including the freedom to fail during the experimental stage of scientific development.

The tainted blood crisis illustrates that failure can come through ignoring scientific evidence and through negligence. The trials of cyclosporine and interferon illustrate that even during pandemic outbreaks we need to recognize that pressing needs do not supersede the need to follow the scientific process. When the stakes are high, short cuts are risky and can result in harm to patients, to reputations, and to trust in the health system at large. Stab-in-the-dark processes drain resources away from other options and rarely succeed. That said, much is to be learned from failures, in establishing a scientific foundation for innovation.

About the Author
by Dr. Neil Simonsen, B.Sc., M.D., D.T.M.&H., A.B.I.M., F.R.C.P.C.

Dr. Frank A. Plummer, O.C., O.M., M.D., F.R.C.P.C., F.R.S., L.L.D., D.R.Sc., is the 2016 Canada Gairdner Wightman Award winner. The selection committee cited his ground-breaking HIV/AIDS work in Africa and his leadership at the Canadian National Microbiology Laboratory, where he led a team of researchers and public health practitioners combating infectious disease epidemics at home and abroad; notably, it cited the global leadership role he played in combating the SARS, H1N1, and Ebola pandemics.

Dr. Plummer's work on HIV/AIDS, a baffling disease in the early 1980s, led to a better understanding of the risk factors for its spread and transmission. His team of researchers made four key discoveries that have saved millions of lives globally. The first discovery was to describe the continental spread of HIV/AIDS in Africa and to note that this transmission involved heterosexual women (female-to-male transmission); the second discovery was that HIV could be transmitted via breast milk from mother to child; the third discovery was that circumcision protects men against acquiring HIV; and the fourth discovery — and the most far reaching — was identifying a group of women (sex workers in the slums of Nairobi, Kenya) who have what appears to be natural immunity to HIV/AIDS.

Dr. Plummer's Canada Gairdner Wightman Award is significant beyond the fact that he certainly merits it. Although the Gairdners are not as well known as Nobel Prizes within Canada and internationally, they have come to be nicknamed the "mini-Nobels" because many Canada Gairdner recipients have later gone on to win a Nobel Prize. In fact, since the Gairdners started in 1959, eighty-three Gairdner winners have also earned Nobel Prizes. His Canada Gairdner Wightman Award is all the more impressive because, just like the Nobels, the Canada Gairdner Awards have no eligibility exclusions based on nominees' nationality. Simply put, Dr. Plummer ranks among the best in the world.

Under his guidance, the National Microbiology Laboratory developed into one of the world's premier institutions in the research, detection, and response to global infectious disease and biosecurity threats.

Dr. Plummer has a stellar record of accomplishment publishing ground grounding research, with close to four hundred peer-reviewed articles in prominent scientific journals and key papers highly cited by his peers.

A Distinguished Professor at the University of Manitoba, Dr. Frank Plummer was a Senior Adviser at the Public Health Agency of Canada. He is an Officer of the Order of Canada, which is Canada's highest civilian honour, and a recipient of the Order of Manitoba. In addition to his recent Canada Gairdner Wightman Award, he has received a Killam Prize, Prix Galien, the McLaughlin Medal from the Royal Society, the Scopus Award from the Hebrew University of Jerusalem, the founding Grand Challenges in Global Health Award from the Bill & Melinda Gates Foundation, the I.S. Ravdin Basic Science Award from the American College of Surgeons, "Researcher of the Year" from the Canadian Institutes of Health Research, the Malcolm Brown Award, two honorary doctorates, and many other national and international awards and honours. He serves on numerous international and national boards and committees.

Born in Winnipeg, Dr. Plummer graduated as a medical doctor from the University of Manitoba and, after completing his clinical training, travelled to Nairobi, Kenya, in 1981 on a one-year fellowship. He returned to Nairobi in 1984 to develop a research program on sexually transmitted infections and subsequently spearhead the development of the Kenya AIDS Control Program established by the Universities of Manitoba and Nairobi. Praised internationally for its excellence, the program has prevented millions of HIV cases around the world. As one of the world's leading researchers in infectious diseases, through his work in the field, he has expanded overall knowledge and understanding of infectious diseases and changed public health policy in Canada and globally.

In 2000 Dr. Plummer bid adieu to Africa to return to Canada to run the federal government's National Microbiology Laboratory, which, at the time, faced considerable challenges as an institution, including recruitment and scientific credibility.

7

NAUSHEEN SADIQ
Postdoctoral fellow, McGill University

To Fail and Tell the Tale — A Story of Science

I met Nausheen in 2015 at an event in the French ambassador's residence in Ottawa, honouring the L'Oréal-UNESCO For Women in Science Award winners from that year. Nausheen was, at the time, a Queen's University Ph.D. student in Analytical Chemistry. She is currently a postdoctoral fellow at McGill University and spends time back home in Calgary and Pakistan. The evening, as the winners took turns speaking on the stage, was a very traditional awards experience — until Nausheen got to the microphone. Immediately she captivated the audience by laughing at herself and making fun of her research. She was genuine, she exuded confidence, and she made science understandable to everyone in the audience. I was hooked instantly. Those qualities are the reasons I asked her to participate in this project — she speaks from the heart and will inevitably continue to be an amazing Canadian. Keep an eye out for her!
— Alex Benay

What Is Success?

What is success? This is an age-old question and one that is often defined by the individual or individuals answering the question itself. Is it keeping your New Year's resolutions? Is it successfully conducting an experiment without breaking any glassware? Or is it winning a

Nobel Prize? However success may be defined, in the realm of science it likely begins with a successful experiment and builds to the point where a Nobel Prize is won!

While a Nobel Prize is not awarded to everyone, success in science has a wide range. Depending on the type of science you do, success can be many different things. It could be developing a drug that can save countless lives. It could be helping an animal in danger avoid endangerment or even extinction. It could be further understanding what is smaller than an electron or even understanding what is in our food and how to make it safer for human consumption. Whatever your field or whatever the success, it doesn't have to be life altering. What it needs to be is continuous and relentless. When we stop moving forward, we do not become stagnant; we actually take several steps back. The world around us is constantly changing, and a scientist's job is not only to ask questions that the world wants answered, but also the ones it has yet to ask.

Success may have a plethora of definitions, but I can tell you it is a nonstop struggle. The reason for this is that success does not mean your work is done; rather, it is usually an indication that the work has just begun. Good luck!

What Is Failure?

So then, what is failure? Does failure hit on Blue Monday in January when your resolutions seem too difficult to achieve? Is failure breaking glassware every time you conduct an experiment? Or is it the fear that you will never win a Nobel Prize?

While failure is typically the opposite of success, in science, failure often seems to be going back on your word. Especially when you have convinced enough people that something is true.

For example, when I was in grade five, I made a presentation on my favourite planet: Pluto! I had a whole skit prepared where I emphasized how far away Pluto was from us here on Earth, and yet how similar it was with its one moon (Charon) — which we now know is just one of many! I got a very good mark, but looking back I now realize I had taught my class an untruth. For in fact, everything I knew about Pluto being the final planet in our solar system was a mistake. In 2006 we were told that Pluto was no longer considered a planet, but rather a dwarf planet, and our solar system had only eight planets. By then I was far past grade five. But I thought back to those long-ago days when I realized that everything I thought I knew to be true had begun to change. What else were we wrong about? What else would change in my life that I had believed to be fact? Had science failed me? Had NASA failed me? What was next?

Now let's take it way back. I mean *wayyyyyy* back. Back to when the world was flat. What an interesting time to be alive! Pluto was a shock to us because we had literally seen it. We knew better, but somehow were still wrong, and we had the full support of twenty-first-century technology on our side, unlike folks living at

A new look at Pluto, Charon, and the four other moons we (NASA) recently "discovered."

the time the world was thought to be flat. Now, from my understanding (I wasn't actually around), when the theory that the world was flat was popular, folks thought that if you explored a bit too much out there, you would literally fall off the face of the earth. That sounds scary. Being a woman of science, I am not sure what I would believe. If someone told me the way I would die was by falling off the edge of the world, I wouldn't love it, to be completely honest. So, would that stop any of us today from trying to find out if it was true? Luckily, there were brave individuals who tried to see what would happen. They ventured beyond a certain point where they expected to face doom, but they did not perish. They kept going.

This is not the stand-alone example of past theories. It was also once believed the earth was the centre point of the universe — which we also now know to be untrue.

So again the question becomes, How is this all failure? I would suggest that this all becomes a failure not necessarily because we were wrong in what we thought to be true but rather because we were unable to conceptualize that there could be an alternative. Failure is not always being wrong, but being unwilling to accept anything other than your theory. We might also consider it a failure when you have misled millions and more to believe something that is just plain wrong.

At the end of the day, failure and success are intimately related — you often can't have one without the other. Let's take a journey together where we are happy to fail because it means success is right around the corner!

What Is Science?

So, what *is* science? How is it defined? If you haven't done so, please take a look at a dictionary definition. Dictionary definitions aren't great. They're vague. After reading several I would never identify myself as someone who studies *dictionary-defined* science. As a scientist I am a person who has spent my life learning about sciences that do not fit these unfulfilling descriptions. There are many ways beyond the dictionaries to illustrate science. Some would see science as the exploration of the unknown, others as something that is divided into "hard" and "soft" elements. In schools, they often split the sciences into chemistry, physics, and biology. The benefit of this is that it allows students to break the sciences into more graspable pieces of a very complex puzzle.

However science may be defined, at its core it is an exploration of the unknown. Scientists never stop with what is visible, known, or within reach. Science is taking what you know and finding out more. My favourite part about being a scientist is not creating something from scratch, it is taking the work done by those who came before me and building upon it. It is asking questions — sometimes questions we never knew needed to be answered. At times science is finding out that the answer to the question is something completely different that we had anticipated. Our goal is to never stop asking questions or looking for answers.

There will always be unknowns. A scientist's job is to explore those unknowns and provide people with the knowledge and power to make the most of the results obtained.

Science in the Classroom

My parents always taught me that teachers were my second parents. I was to respect and care for them as such. Teachers are with you a lot in your formative years. They have the ability to shape your life and help you reach your full potential. Yet one of my biggest issues then was how science was taught.

When I was in junior high school we started learning more and more about what science was. We were building the foundation that would then lead to our high-school science studies. In Alberta, our science classes were split into the following:

Grade ten:	Science 10 – a combination of biology, chemistry, and physics
Grade eleven:	Biology 20, Chemistry 20, and Physics 20
Grade twelve:	Biology 30, Chemistry 30, and Physics 30

By grade eleven we were getting into the nitty gritty of each of the three subjects. I obviously enjoyed the course material that I was taught, as I went on to get a Ph.D. in chemistry. One thing always stood out, though. As we went from year to year and grade to grade, we kept being told "Wait! There's more!" I hated this! How was it possible that every year we would learn the same concept, but suddenly there was more to it? We were told that something was correct and forced to understand the concepts, and the following year we found out it was all a simplified version. I understand there is an accepted way to teach students, as there is only so much material that can be fit into a week, a semester, or a year. However, for the students who are able and inclined to further their knowledge, is this fair? Are we robbing our future scientists in their pursuit of knowledge by telling them this is all there is? If a student were to challenge the status quo, read material outside the classroom and know better, would the teachers still stick with the curriculum for that student? If we were told that the sun went around the earth in order to explain things better in the short term, would we not be expected to lose faith in the system the following session when it was exposed as a lie?

My issue is not with the system itself. I understand that only so much information can be given to a group of students at one time. But I think that if we expect youth to ask questions and challenge authority, we must give them the freedom they need when it comes to what they are taught in the classroom.

My favourite example is from chemistry class. The ideal gas law ($PV = nRT$), pronounced "pivnert" (at least, that's how I pronounce it), is something I was misled about from high school until well into my university years. In high school the equation was simple. Nothing too complex, and the units could change based on the question given. The ideal gas law is a combination of Boyle's law, Charles's law, and Avogadro's law. In an ideal scenario, this equation is gold! But things get interesting when conditions aren't ideal. I took an advanced placement chemistry course in grade twelve, and suddenly we were introduced to the Van der Waals equation. It is an extension of the ideal gas law for when conditions are far from ideal. (They are always far from ideal.) Suddenly everything I had learned for years seemed like a sham. The Van der Waals equation was all well and good until I started first-year chemistry at Queen's University. I felt confident that I had the tools I needed to succeed, and part of that was my knowledge from the year before. Yet at Queen's they proceeded to teach us the ideal gas law as if it was true. I had gone back in time. I knew better but I acted as if this was all I knew. Then in upper-year courses where we learned about "non-ideal" behaviour, suddenly it was "déjà vu all over again." I'm lucky that my love for chemistry allowed me to look past the deception.

I think we need to explore learning outside the classroom, which brings me to the second flaw in teaching. We are shifting toward online learning and online classrooms, and while these do provide opportunities like never before, they still have their shortcomings. While the ability to connect instantly around the world is invaluable, the issue arises when students are expected to put in several hours of "self-exploration" with provided resources. I remember at times it was so difficult to stay on top of classes, labs, midterms, and finals that finding the time to read additional content was a luxury. As technology advances, we need to make sure we are realistic in the way we communicate information and ensure that teaching is still done with the student in mind.

The Overqualified Job Market

To get a job and live to tell the tale! I'd like to tell you the story of a gentleman I met the other day. He was helping to set me up at the local gym, where he worked (the whole New Year, new me mentality was still alive) and asked how long I would be in town. I replied that I wasn't staying long and that I would be returning to university for a few more months as a research associate after my Ph.D. He told me that he had done a Ph.D. in physiology and then gone on to three postdoctoral fellowships in both Canada and the United States. Perfect! My heart sank a little. This man in front of me was working at a gym instead of sharing hard-won knowledge and resources with the academic or industrial world, and not because he didn't want to, but because he couldn't.

We have a large group of highly qualified graduates saturating the job market. On top of that, we don't have jobs available. In academia, the problem arises when professors with tenure are able to stick around and the new talent must wait them out. We are producing far more graduates than there are employable positions. We need to be proactive. We know how many students are being enrolled, so we have an idea of how many students can be expected to graduate every year. We need to ensure that people don't pursue higher education only to move back into their parents' house after they get their M.Sc.s and Ph.D.s. As much as Mom and Dad love us, I am sure they want us to get a job and move out — and visit on holidays.

The job market crisis is hitting all Canadians. This isn't a problem simply for the sciences. It is one that we need to address and work to resolve. It is hard to convince someone to pursue higher education and take on more debt if it will not lead to a brighter future in the long run. Education is an investment in one's future, and that is something that can never be taken away — but the pursuit of knowledge should be rewarded with suitable employment.

Scientific Communication

When you are working toward a graduate degree, when you work in a research lab or even when you have your own lab, the goal is always to *publish*. Publishing is one of the most important parts of academia, as it allows you to share your research with others in your field, regardless of where they are in the world. The more well known the journal, the harder it is to publish in it. The goal, then, is not only to publish but to publish in the most prestigious journal possible.

The race to publish often leads to problems. In an effort to be the first to publish original content, some researchers publish papers that do not actually show the best method or the most efficient way to conduct analysis; rather, they show that these scholars were the quickest to get their work reviewed and published. In my experience, I have very often found papers that don't fully explain the procedure so it becomes difficult to repeat the analysis independently.

One of the cornerstones of publishing in journals is the peer-review process. This means a group of peers in your field will anonymously review your submitted manuscript and will submit requests for revisions; they will also comment on whether they deem the manuscript's findings to be valid. If the reviewers deem your manuscript acceptable, it can be published with minimal revisions. If they find issues, it will be sent back for major revisions, and if they still deem it unacceptable it can be rejected. When qualified reviewers cannot be found, journals are not prompt in replying to submitters. In some cases the only qualified reviewers may be at work on an identical project and are trying to publish the findings themselves. Obviously, in this scenario a conflict of interest exists, and they should not take part in the revision process. Reviewers do not always declare conflict, though. The other issue with having peers review your work is that there is no standard of review. Some peers may make the publication process harder than it needs to be, and others may make it too easy. Although the peer-review process has clear benefits, we need to make sure there is a more transparent and efficient means of review.

Even though the goal is to publish as much as possible, this doesn't mean publishing is a quick process. Keeping in mind the many journals that are released every year and how many submissions each must receive every day, it is no surprise that publishing is a time-consuming process. There have been times that the date a paper was submitted and the date the paper was published were a year apart. That means that a project that was deemed original and of importance still took a year to be made readily available to the public. Beyond a thorough revision process, the time to publish is far too long. We need to make an effort to get high-quality and relevant papers published ASAP!

The goal for good research is to be able to share findings with anyone who has the means to access the research. When scientific writing is so highly specialized that you need deep knowledge of the field to understand it, it becomes a problem. If we want our work to be accessible to the world, we have to present it in a format that is understandable by all. This is an area where I feel there is great room for improvement, and I admit I am guilty of the very same failure to communicate. Throughout my degree, I learned to explain my research in the same way to someone who has no knowledge of chemistry as to someone who is an expert in inductively coupled plasma–mass spectrometry (ICP-MS). We need to teach our scientists not only to write well and write scientifically, but also to write accessibly.

Failures in the Lab

There are countless failures in the lab. Failure might be an unsuccessful experiment; you may fail to have your work published in a peer-reviewed journal or even fail to win a grant to fund your research.

Failure in the lab is not restricted to the undergraduate student, Ph.D. candidate, or even tenured professor. Failure is part of the day-to-day function in a lab; however, we do not always call it that. We find a way to justify our shortcomings into excuses as to why failure may have happened. The failure of a simple experiment could be easily blamed on the instruments used, the failure in publishing could be that the reviewers were too picky ... and finally, there simply aren't enough grants for all the great work that is being done! However you look at it, failure in the lab can always be justified (and that helps us thrive). There are, however, times in the lab when failure cannot be overlooked. These are moments when collaborations between institutions, departments, and even within individual labs develop issues. Sometimes the researchers are ready to work but their instruments are broken or require repairs that the lab simply cannot afford. The biggest failure in the laboratory in Canadian science is, in my opinion, this situation, when individuals are willing to work but do not have the means, the resources, or the support to do so.

I will give you a personal example. I will leave some details out in order to protect as many parties as possible, but I think it is still a story worth sharing. I was honoured to receive several scholarships for travel to a foreign country to further develop my Ph.D. research and collaborations. At the time, we were told that state-of-the-art instruments were being ordered for my arrival. Part of my role would be to train and assist the entire team in this foreign institution in using the instruments. I was also to use the time available to do several experiments and hopefully publish several papers!

It was only after I had embarked on my adventure and travelled to the foreign country that I learned that not only were these state-of-the-art instruments not available, but they had yet to be ordered. It is safe to say that I was less than impressed. I was in a foreign land where I didn't speak the language and I felt that I had been tricked. Despite this, I made the most of the experience, worked on projects that I had never imagined, and learned a great deal. The issue wasn't that I didn't enjoy the experience or gain valuable knowledge and resources, but rather that I didn't achieve the goal I had set out to achieve. As an update, the instruments have still not arrived, a year after my return.

This was likely the greatest lab failure in my research. I had set out to accomplish great goals, but due to lack of clear communication, bureaucracy, and time constraints, I was unable to accomplish what I had set out to do. Don't get me wrong, I'm not saying that the entire trip was a failure: in that same country I also visited a different institution and was more productive than I could have ever imagined! If I am lucky, I will get to write about that in a book of Canadian *successes*. (The thing is, we are Canadian ... gloating is not our strong suit, and so that book may never come to be!)

Women in Science

Let's talk numbers. The L'Oréal Foundation #ChangeTheNumbers campaign addresses the number of women in science. "Only 30% of researchers are women. Only 3% of Nobel Prizes in the sciences have been awarded to women. Only 11% of the highest-level academic scientific positions are held by women." So the question is: Why so few? When increasingly more women are attending university than men, why is there still such a discrepancy in the sciences?

There are now many groups that encourage young girls to get engaged in "STEM" (science, technology, engineering, and mathematics). When I was younger, I didn't have these opportunities. I think these are fantastic initiatives, and I look forward to continuing to work with groups like the Canadian Association for Girls in Science, and Girls Inc. I have a Ph.D. in chemistry and I credit many people who helped me get there, but most of the credit I give to my teachers. To those who sparked my interest in the sciences, thank you.

We encourage a discrepancy between the sexes early in life. We live in a world where girls are taught to play with dolls and boys are taught not to be afraid to get their hands dirty. When I was growing up and I wanted to play with the toys my older brother had, I wasn't praised as a young scientist or encouraged to explore with my curious mind, I was labelled a "tomboy." My lack of interest in the colour pink or in Barbies was seen as a shortcoming. It is not that I didn't have the opportunity to play with a mix of toys. My Easy Bake Oven got its fair share of use, and is likely why I won home economics awards in junior high school, but I also liked to get my hands dirty. Even in 2017 pink is still a girls' colour and blue is still for boys. We need to work to change this, and one way is encouraging girls to explore the sciences from a young age. All children should be inspired by those around them to do what they love.

Award ceremony for women in science held by the L'Oréal Foundation in Ottawa, 2015. Nausheen Sadiq is on the far left. May this tradition be carried on by L'Oréal and other organizations into the future!

What a noble pursuit! What could be better than encouraging women and girls to take an active role in the sciences? I think too many of these groups across the country, all with nearly the same mission statements, have been working separately. Their lack of collaboration makes their noble pursuit something of a failure. I have volunteered for some of these groups. It turns out that when you attempt to collaborate and draw similar groups together, it rarely works. If we all have the same mission statement and the same end goal, why is it then such a difficult task? Let's come together! Let's put a truck in one hand and a doll in the other hand, anything that shows that a girl can make her own choices. Let's work to help girls to explore the sciences and also teach our boys that they should expect strong competition from their female counterparts. Let's make sure that when we push our girls to love science we push our boys, too, because the future of science is bright in this country, and we need everyone to take an active role!

The tides are changing in Canada and across the world. I have faith that the contributions women make to science and the world in general are just beginning. We are on the verge of a breakthrough, and I just can't wait to see what comes next.

Where Did We Go Wrong?

Is there something we are doing wrong? Was there something we could have done to prevent these mistakes? Is failure inherent in our society, our country, or our sciences? My answer is no: failure is a part of life. Any system that is put in place will have its flaws. The key is to know where we are going wrong, to be able to reflect on this knowledge, and to realize that it cannot be ignored. We only truly fail when we see the errors in our operations and do not work to fix them. If we follow the wrong path we get nowhere. In the sciences, as in all fields, we need to be moving constantly forward. The sciences in

Canada need to reflect on the past, work to perfect the present, and plan to thrive in the future.

Conclusion — Strength in Failure

There is unacknowledged strength in failing. It is after truly failing that we can pick ourselves up, brush off the dirt, and get back to business. There is always room for improvement. This still holds true in Canadian science, be it in the lab, with collaborations, or even in our philanthropic efforts. By becoming aware of our shortcomings, we are able to grow stronger. The future for Canadian science is bright. We are starting to invest in our scientists again, and we now have the technology and equipment we could only dream of in past years. As our children ask questions, and continue to ask questions, we need not only to answer as best we can what they ask now, but plan to answer what they will ask in the future. As we explore the sky above us, the ground below, and everything in between, I do not doubt that our failures are just the start of Canada becoming one of the leading nations in science.

8

ROBERT THIRSK

Former Canadian Space Agency astronaut

Preparation for Failure

I first met Robert for lunch, during which he proceeded to give me an ear-ful on the shortcomings of the Canada Science and Technology Museum in Ottawa. He was right in everything he said that day. The museum was due for a major overhaul, and that's exactly what happened. We tore down the walls, put them back up, and reintroduced 80,000 square feet of new exhibitions from scratch, all within two years — something that had never been done before. In the process, I gained a friend and the museum gained an ally. Robert is one of the most accomplished human beings I know. Starting with the obvious: he's an astronaut. How amaz-ing is that? He's also Chancellor of the University of Calgary, he has a few postgraduate degrees and, even more importantly, he is a man of passion who agreed to share his perspective on failure with us in the hope we can all learn from his approach. Thank you, Robert!

— Alex Benay

"*T*en ... *nine* ... *eight* ... *seven* ..."

Six seconds before liftoff, the space shuttle's primary com-puters send commands to the three main engines to ignite. These liquid-fuelled engines start with a deafening roar. Over the next three seconds, the computers verify that each engine reaches at least 90 percent of its rated thrust. At 0.3 seconds before liftoff, the

Thirty-one million newtons of thrust launches the space shuttle skyward. For those of us onboard, it is a wild, wild ride.

computers send a command to the two solid rocket boosters to start up. These powerful boosters instantly reach full thrust.

At T=0, the eight large hold-down bolts that secure the space shuttle onto the launch pad are explosively released. Freed now from the launch pad, the shuttle — including the orbiter *Endeavour*, its large orange external tank and two solid rocket boosters — leaps skyward in a fury of smoke and flame. I am jolted forward in my seat. Liftoff!

Riding aboard the space shuttle as it thunders away from its Florida launch site with all five engines at full power is an unforgettable experience. Clearing the launch tower, the rocket engines are burning propellant at a rate of eleven tons per second. Although securely strapped into our seats, my crewmates and I experience bone-rattling vibration.

Kevin Kregel, our shuttle commander, is seated forward in *Endeavour*'s flight deck and to my left. Pam Melroy, our pilot, is forward and to my right. Kevin and Pam are both former U.S. Air Force pilots and guide the shuttle during its ascent. Surrounding Kevin and

Ascent is a dynamic phase of flight when the shuttle crew needs to be especially vigilant.

Pam are panels with hundreds of displays and flight controls — hand controllers, computer monitors, keyboards, meters, switches, circuit breakers. The instruments enable them to monitor the shuttle's status and control its performance during flight.

I sit in the centre seat of the flight deck. My role as flight engineer is to keep the crew on the timeline and call out any procedural actions to be taken. I hold checklists and a notebook. Tethered nearby are my cue cards and reference books.

Seven seconds after liftoff, the shuttle clears the launch tower and begins a roll around its longitudinal axis to align with our desired orbital plane, putting us into a slight heads-down, wings-level orientation. Kevin monitors a successful roll on his flight instruments, keys his mike, and reports, "Houston, *Endeavour*, roll program." The Capcom at the Mission Control Center in Houston responds on the air-to-ground radio loop, "Roger roll, *Endeavour*."

During ascent my crew is focused. Conversations among ourselves and with Mission Control are concise, and peppered with acronyms and jargon. There is no time for chit-chat during this short-duration, highly dynamic phase of flight. We are aboard the most complex space vehicle ever built and sitting atop two million kilograms of explosive propellant.

Thirty seconds after liftoff, Pam scans her computer displays and verifies engine thrust levels dropping from a nominal 104 percent down to 67 percent. She reports, "Throttling, three at 67."

This down-throttling of the three main engines is expected. To minimize the aerodynamic stresses on the outside surface of the shuttle, the engines are automatically throttled back as we go transonic. A half minute later the engines rev back up to full power, and Pam states, "Throttling, three at 104."

Mission Control calls, "*Endeavour*, Houston, you are go at throttle up."

"Roger, go at throttle up," Kevin replies. We are now a minute into the flight, supersonic and climbing through an altitude of ten kilometres.

"Beeep." A warning tone sounds. I wince. Scanning my computer display, I notify my crewmates, "I see the 'APU Speed Low' message for APU-2."

Pam looks at her display and adds, "This could be a sensor problem or an RPC failure. My data shows the APU is still running."

An APU is an auxiliary power unit. It pressurizes an associated hydraulic system. The hydraulic system, in turn, drives the gimbals of the rocket nozzles that are currently steering our shuttle on its precise path to orbit. At the end of the mission when we fly home, the hydraulics will power the orbiter's elevons, speed brake, and rudder. On final approach to the runway, they will be needed again to deploy the landing gear.

For redundancy reasons, the orbiter has three independent APUs — all located in *Endeavour*'s aft compartment near the main engines. The brilliant engineers who designed the shuttle back in the 1970s ensured that, wherever possible, systems were built to a "fail operational, fail safe" standard. This means that if a critical system (such as an auxiliary power unit) should fail, then a backup system would allow full functionality. And if that backup system should subsequently fail, we may lose function but a third system would at least allow us to complete the mission safely.

Mission Control calls with an update, "RPC Bravo for APU-2 controller is down. No crew action. The APU is still powered."

There are hundreds of RPCs (remote power controllers) aboard the shuttle. They function like circuit breakers. If there is a short circuit in an electrical bus, an RPC will trip open to unpower its circuit and protect the downstream components. In our case, an RPC that provides power to the controller for APU-2 has just tripped open.

The good news is that this controller has a second power source (thank you, engineers!). APU-2 continues to function well and is providing hydraulic power to the nozzle gimbals.

However, a few seconds later Mission Control alerts us to a different problem: "We are monitoring a pressure drop on hydraulic system three. No crew action."

Yikes! It seems that we are now developing trouble in another hydraulic system. The dropping pressure that Mission Control reports is concerning and may indicate a leak in system number three. I log these malfunctions in my notebook.

Glancing out the forward flight deck windows, I notice the colour of the sky transitioning from blue to deep purple as we rapidly gain altitude. Two minutes into the flight and at an altitude of forty-five kilometres, the two solid rocket boosters have used up all their propellant and are jettisoned. Kevin quips, "Good riddance!" I smile. The SRB rockets are so incredibly powerful that they are worrisome. Now that the SRBs are gone, the noise level in the flight deck drops. We race upward on the remaining three main engines — a noticeably smoother ride.

I peer over my crewmates' shoulders at the meters on the forward cockpit panels to monitor our speed. As we approach a velocity of Mach 5 (five times the speed of sound), I call out "Two-Engine TAL."

TAL means transoceanic abort landing. It is one of the options we could take to end our mission if something should go wrong with our flight. If, for example, a shuttle main engine were to fail in the next moments, then we would not be able to attain the speed that is necessary to reach orbit. We would instead need to abort the mission and fly back to Earth. My call "Two-Engine TAL" indicates the point in our ascent at which our crew could consider performing a landing in western Europe if we were to lose an engine. However, in reality, all three main engines continue to perform well and we're on our way to space.

A few seconds later Mission Control repeats the same "Two-Engine TAL" call. Hearing Capcom's voice reassures me that we have a good communication channel and that the ground is closely tracking our ascent. The Houston-based flight controllers have additional data and insight about *Endeavour's* status that would be valuable to us if we were to run into a problem.

"Beeep." Another warning tone sounds. Scanning my display, I see failure messages for several equipment items. I am dismayed. I consult my reference book and then inform my crewmates, "Main B APC-5 has failed. We've now lost APU-2."

Aft power controller number five, a sub-bus of the main B electrical bus, had been the other power supply for auxiliary power unit number two. With the loss of this redundancy, APU-2 is now unpowered. Rats! Pam performs an orderly shutdown of APU-2, and I assist her with the procedure steps.

We only need one operating APU to complete the mission. While we have lost APU-2, APU-1 is still performing well, and APU-3 with its slow leak is somewhat functional.

A moment later, I call out "Negative Return" as I see our velocity approach Mach 8. *Endeavour* now has too much altitude and velocity to successfully turn around and return to our launch site in Florida if an abort landing were to be required.

"Pam, I didn't hear the ground repeat the Negative Return call. Do we have signal strength?" I ask.

Pam glances at her radio communication meter and replies, "No, we don't." She calls the ground, "Houston, *Endeavour*, communication check." After several seconds of silence, there is no response. "Crewmates, it looks like we have a comm failure with the ground. We're on our own. I'm working the Comm Lost procedure."

Without the voice calls from Mission Control, I need to be extra vigilant as flight engineer — monitoring our velocity, abort boundaries, my cue cards, and any system failures.

"Beeep." Another warning tone. Aargh, will it never end? I look up at the forward panel and see that the "APU Overspeed" warning light and "Hydraulic Pressure" light are lit. I break into a cold sweat. It looks like APU-1 has just failed due to an overspeed. This is not a good day.

Kevin calls out, "Two hydraulic systems down, and the third one failing. Bob, what do the flight rules say about our situation?"

I'm sure that Kevin knows the answer but he wants to hear confirmation from me. The shuttle's three main engines are performing well and we have enough power to get to orbit. Nevertheless, we must abort our mission and return to the ground as quickly as possible. In addition to being a rocket, the orbiter is also an aircraft. At least one working hydraulic system will be needed to drive *Endeavour*'s aerosurfaces for the descent, approach, and landing when we return to Earth. With two hydraulic systems having failed and a third one leaking, we could have a complete loss of hydraulics at any moment. We urgently need to get to the ground.

"We're not going to space today, Kevin," I respond. "According to the System Flight Rules, we need to abort. Select TAL."

Pam adds, "My data shows enough HYD-3 pressure to support a landing."

"Concur," says Kevin. "We've exceeded the Negative Return boundary velocity, so I'm selecting TAL abort."

The flight controllers at Mission Control in Houston oversee every aspect of shuttle flight and are ready to assist the crew.

Since *Endeavour* is now unable to return to Florida, our next best option is a transoceanic abort landing. A TAL abort puts the shuttle into a suborbital trajectory — a giant hop across the Atlantic Ocean. A TAL is theoretically designed to provide an intact landing of the orbiter on a predesignated runway in Europe or Africa. I say "theoretically" because, in fact, a shuttle abort landing has never been performed. I now feel my heart pounding in my chest.

Kevin selects the abort mode by turning a rotary switch on the panel in front of him to the TAL position and then attempts to engage it by pressing a pushbutton. "No joy," our commander calls out. "The rotary switch has failed."

"Call up SPEC 51," I interject. "Enter ITEM 2 EXEC to select TAL, then ITEM 3 EXEC to initiate."

Kevin calls up the SPEC 51 override display and keys in the two commands. The backup method works as the TAL trajectory pops up on his monitor. Phew! I record our velocity (Mach 10) and altitude (120 kilometres) in my notebook.

With the abort mode now selected, a series of pre-programmed events are initiated. The three engines need to be throttled back from 104 percent to 67 percent. This should happen automatically, but due to the pre-existing hydraulics failures, throttling for two of the engines is disabled. Pam instead throttles them back manually.

Our new objective is a controlled re-entry and the safe return of the orbiter and our crew to ground. Everyone gets busy. I turn to the abort pages of my checklist, deploy new cue cards, and keep an eye on our velocity and range. Kevin monitors *Endeavour*'s guidance and verifies our landing site. We are targeting a runway at the Zaragoza Air Base in northeast Spain. Zaragoza does not lie directly under our ground track, so with the main engines still burning, the flight computers command the rocket gimbals to steer *Endeavour* closer to the airport. We've lost gimballing capability in one engine, but the other two have enough remaining capability to guide us in the right direction.

Pam monitors the dumping of Orbital Maneuvering System propellant. By dumping prop, she'll reduce *Endeavour*'s weight and reposition its centre of gravity. This will improve aerodynamic control during our gliding descent and landing in Spain.

Glancing out the forward windows I note the sky is now inky black with a few stars apparent — we're on the edge of space.

At Mach 15, the shuttle rolls 180 degrees to a heads-up orientation to ensure a proper attitude for separation of the external tank. The three main engines shut down shortly after and we are all flung forward in our seat harnesses as the vehicle decelerates. Seconds later, the external tank is jettisoned. Thruster jets fire like popcorn to maneuver *Endeavour* away from the tank and we begin a ballistic fall back to Earth. Zaragoza is still five thousand kilometres away but I sense our rapid speed and precipitous descent. We'll be on the ground shortly.

Kevin and Pam have been occupied for several minutes flying the orbiter and preparing for its landing. As I see the approaching coastline of Spain through the forward windows, I must update them on one more significant issue. With the loss of hydraulic systems number one and number two, we don't have the capability to deploy our main gear and nose wheel for landing. Not pleasing news from me.

"We'll do a pyro deploy of the gear," Kevin replies. "Let's hope it works." The pyrotechnics will be our last means to get the landing gear down.

Four hundred kilometres from Zaragoza, we once again hear voice calls from Mission Control and our instruments begin to display tactical air navigation data. While our craft slices downward through the sky like a brick glider, Kevin takes manual control of the orbiter as our speed transitions to subsonic. The 3,700-metre-long runway at the Zaragoza Air Base comes into view. At one hundred metres altitude, Pam arms and fires the landing gear deployment. I whisper a quick prayer, "Oh God, please deploy the gear." Boom! Like howitzers, the pyros fire. Seconds later, Pam reports, "Gear down!" (Thank you, God.)

Kevin brings *Endeavour* down smoothly at a touchdown speed of 360 kilometres per hour — as soft a landing as a commercial airliner. Pam immediately deploys the braking parachute. We've lost nose wheel steering with the hydraulics failures, so Kevin guides *Endeavour* down the runway with differential pressure on the brake pedals.

As we roll to a stop, I exhale, relieved that our vehicle is intact and my crew is safe.

Except for the quiet whir of the cabin fans, there is silence on the flight deck. Nobody talks. A bead of sweat runs down my brow and drops onto my checklist. Exhausted, I look at my watch. Forty minutes have elapsed from launch to landing. That must've been the fastest transatlantic trip anyone ever took.

The silence is broken a moment later by a cheery voice heard from the overhead speaker. "Nice job, crew." It is the voice of our shuttle instructor, Linda Pope. "That was the last of four ascent runs for this evening. Let's debrief them now. Grab a coffee and meet me at the instructor station in five minutes."

My crewmates and I are not in Spain, but in Houston at the NASA Johnson Space Center. The ascent and landing that we just flew aboard *Endeavour* was a simulation.

o o o

Space flight is an enormous undertaking that doesn't happen without considerable planning. The skills of thousands of multidisciplinary specialists and the resources of many organizations are required to prepare a spacecraft and its payload for flight.

Preparation of an astronaut crew for flight is equally daunting. Once in space, astronauts are expected to perform all assigned tasks correctly and quickly. Due to the speed of the spacecraft, the constraints of orbital dynamics, or the tight mission timeline, an astronaut often has only one chance to perform a given task. We need to get it right the first time.

The key to astronaut success is abundant and thorough training. Basic training begins when a newly recruited astronaut joins a space agency. Recruits gain extensive knowledge — in addition to skills and attitudes — unique to the profession. We acquire technical knowledge of spacecraft systems (e.g., propulsion, life support), operational skills (e.g., space walking, robotics) and behaviours (e.g., teamwork, leadership). Training during the first few years of an astronaut's career is like drinking from a firehose — massive amounts of new information must be continually learned.

In later years, advanced astronaut training focuses on learning the all-encompassing skills to competently perform a mission. Working in space when a flight is going well is straightforward. However, there

Crew operations in space can get complicated. It pays to be well prepared.

has never been a mission that has unfolded according to its preflight plan. System software and hardware do not always function as designed (known as "malfunctions"). Unforeseen mission events (known as "contingencies") occasionally occur. The flight timeline, for many missions, is out of date soon after the spacecraft reaches orbit. Since missions can be months long, it is impossible to know in advance what malfunctions and contingency situations might arise during flight. Hence, astronauts need to be ready for anything.

How is this advanced training accomplished? By repeated practice. While astronauts spend hundreds of hours training for nominal days in space, we spend thousands of hours training for possible bad days. Simulators play a substantial role in this phase of training. These simulators are high-tech training facilities designed to replicate aspects of the spaceflight, the spacecraft, or the space environment.

The Shuttle Mission Simulator realistically replicated the controls, data, motions, sights, and sounds of shuttle flight for astronaut trainees — everything except weightlessness.

For example, the Shuttle Mission Simulator (known to astronauts as the SMS) is the name of our highest fidelity simulator. During the shuttle era, it was used extensively to prepare astronauts for ascents and re-entries. Located atop an elevated platform, the SMS operated with motion cues supplied by a six-degree-of-freedom system that simulated all phases of flight, from launch to rendezvous to deorbit to landing. A rotating frame even provided a 90-degree upward tilt of the crew cabin to simulate the acceleration of liftoff and ascent. Until the space shuttle was retired in 2011, SMS simulations (referred to as "sims") honed the operational skills required by crew members to perform competently in space.

Additional simulators of various sizes, shapes, and sophistication provide training for astronauts on other essential skills. The robotics simulator at the Canadian Space Agency in Montreal, for instance, uses virtual reality to model Canada's robotic systems aboard the International Space Station. A large centrifuge at Star City in Russia is used to prepare cosmonauts to manually pilot a Soyuz capsule back to Earth while under g-load. The Neutral Buoyancy Laboratory in Houston is a huge swimming pool used to simulate the weightless condition experienced by astronauts during space walks. The Shuttle Training Aircraft is a corporate jet that was modified by NASA to mimic the instrumentation and flight profile of the shuttle orbiter. Training sessions in this aircraft familiarized pilot astronauts with the unique handling characteristics of the returning shuttle during its steep approach and high-speed landing.

Our collective training experiences in an assortment of simulators, repeatedly rehearsed, prepare us well for all phases of a mission from launch to landing. It's not enough to train until we get it right. We need to train until we can't get it wrong.

In addition to perfecting my operational skills, simulator-based training also helped me develop a unique mindset known as "situational awareness." Situational awareness à la Wayne Gretzky ("skate to where the puck is going, not to where it is") enables astronauts to think ahead and assess evolving trends. Since it is impossible to know with certainty how a space mission will unfold, it is important for astronauts to constantly monitor the progress of a flight.

Astronauts with well-developed situational awareness are most proficient at handling off-nominal situations. They have an uncanny ability to react quickly and appropriately when things suddenly go wrong. Thoughts about the next possible failure preoccupy their minds: what malfunction could happen that would make our current situation even worse? If it were to happen, what should be my immediate reaction?

Each simulation concludes with a debrief. A debriefing session is a lessons-learned exercise that captures what went right or wrong during the sim. It is an opportunity for the crew to consider how we could have better coordinated among ourselves or how we would do things differently next time. The answers to these questions help to sharpen our performance.

The unusual flight characteristics of the shuttle during approach and landing were simulated in the Shuttle Training Aircraft.

For example, simulations held in the Shuttle Mission Simulator facility were directed by a first-rate team of flight instructors. Like Hollywood screenplay writers, our instructors incorporated a variety of contingency situations (e.g., onboard fires, abort landings) and a barrage of carefully staged malfunctions (e.g., loss of a main engine, computer failure) into each simulation script. A well-written script includes a measure of realism, causes each participating crew member to "stretch," and trains the crew to work together as a cohesive unit.

While the crew works through mission scenarios in the simulator, instructors sit nearby at consoles and control the unfolding events by inputting equipment failures and contingencies into the sim at strategic moments. They then observe the crew and evaluate our responses to the off-nominal situations.

The shuttle abort landing simulation that I described at the beginning of this chapter was one of hundreds of training exercises that I have participated in during my astronaut career. Following that evening's four training runs, our instructors debriefed each sim scenario with us. My crewmates and I then reviewed our responses to the scripted failures and described the rationale behind our actions.

There are lessons to be learned by all sim participants. However, that particular evening's debrief was especially beneficial for me. Both Kevin and Pam were experienced shuttle pilots, while I was still early in my training program. I learned, for instance, that a crew should never attempt to restart an APU that has undergone an overspeed failure (doing so could have catastrophic consequences). I also learned about other orbiter system failures (e.g., a large cabin pressure leak or a cooling system failure) that would require aborting the mission quickly during an ascent and landing.

That sim also reinforced to me the importance of the nominal timeline. Even though I was busy dealing with a contingency, I shouldn't get so distracted with the abort procedures that I neglect to also perform the critical nominal steps.

My astronaut career regularly took me to the limits of my physical, mental, and emotional capabilities. For that reason I tolerated mistakes that I made when training at my limits. Furthermore, I regularly sought and valued feedback from my instructors. These dedicated professionals best knew my weaknesses and what specific skills needed improvement. Constructive criticism of my performance by instructors as well as peers enhanced my preparation for flight.

Despite all the training and preflight testing, very few space missions have been perfectly flown. Hardware breaks down and software gets hung up. And every astronaut will admit in hindsight that she or he could have performed a particular task in a more effective manner.

A crew contingency could occur at any time — from launch to landing or anytime in between — and without warning. Due to the harsh environment, high speeds, and huge energies involved in spaceflight, mishaps such as the *Challenger* accident can be unforgiving and catastrophic. In the half century of human spaceflight, eighteen astronauts and cosmonauts have died while serving aboard spacecraft. I lost seven dear friends in the *Columbia* tragedy in 2003. Several more astronauts and cosmonauts have died during ground-based training.

Risk continues to be an inherent part of training and missions. To say that another spaceflight accident involving astronauts won't occur is like saying that there will never be another motor vehicle fatality on the Trans-Canada Highway. Nevertheless, space agencies are not risk averse. We prepare thoroughly and do everything we can to minimize the risk, but we know that someday there will be another loss of life in space and more casualties on the highways.

We don't attempt to eliminate risk. If 100 percent of our spaceflights are successful, then those missions are undoubtedly infrequent and of lesser importance. The only way to eliminate risk in space programs is to keep all rockets on the ground — but that is not what rockets are meant for. Space agencies instead design bold, trail-blazing missions and then manage risk in a rational way.

If I were to die during a spaceflight, I would certainly want my colleagues to investigate the cause of the accident and fix the problem. I would then urge them to return to flight as soon as possible. Human space exploration is an important undertaking with boundless possibilities and societal responsibilities. It must continue.

The space program provides useful spinoffs (economic return, scientific advances, new technologies), but I believe that it is its intangible benefits that bring most value to Canada. The public role I fulfilled as an astronaut has repeatedly shown me that the work we do motivates young people to study science, technology, engineering, and mathematics. It inspires thousands of Canadians to stretch their capabilities and find solutions to the difficult social problems facing our nation.

There are leadership lessons to be learned from human spaceflight's approach to risk management — valuable lessons that can be applied to terrestrial ventures. While the risks of doing business on Earth may not involve life and death situations, the stakes are equally high. Many organizations in Canada are admirably pushing the frontiers of science, technology, and medicine to accomplish something transformative for society. Working in such uncharted business territory means that enterprises on the cutting edge of innovation will periodically encounter crisis situations — crises perhaps associated with product failures, security breaches, or litigious customers. While it is impossible to predict the nature or timing of these crises, they should be expected as part of the normal course of doing business and as inevitably as death and taxes.

My experience as an astronaut has taught me that risk management must be advocated at the highest level and driven throughout the organization. The late U.S. president John F. Kennedy set a good example. Many years ago when the United States was initiating its Apollo Program, Kennedy told Congress, "I believe that this nation should commit itself to achieving the goal, before this decade is out, of landing a man on the moon and returning him safely to the earth."

While JFK's beautifully crafted statement directed his nation to pursue an ambitious vision, it also placed responsibility on his NASA administrators to manage the risks.

It is organizational culture, not administrative process, that underlies successful risk management programs. The chief executive is responsible for establishing a culture that is not risk averse and that encourages novel approaches to problem solving. Calculated risks that advance the organization are supported, while unavoidable risks are managed. Resourceful employees who pursue bold initiatives are recognized and rewarded. Complacency and disregard of safety, on the other hand, are not tolerated.

The development of a risk management plan is only a first step for an organization. The soundness of the plan and the responsiveness of the staff to simulated crises must also be regularly exercised, as is done in human spaceflight. It is naive to think that the contingency response team will be confident and perform well without regular training. It is naive to believe that untested processes will unfold smoothly or address all issues on the day of an actual crisis.

Simulations should be championed by the chief executive; they are the cornerstone of risk management training. Simulations of possible contingencies add credibility to the training and provide a safe environment where mistakes can be made — a very effective way for staff to learn. In such a training environment, failure is indeed an option. Through repeated practice and review of mistakes, the response team members learn their roles, develop an operational mindset, and begin to function as a unit. Thus, on the day when it will really count, the organization will be ready.

In summary, training for spaceflight is not easy; it is the most challenging thing I have done in my life. I again take inspiration from John F. Kennedy when he said, "We choose to go to the moon, not because it is easy, but because it is hard, because that goal will serve to measure the best of our energies and skills."

Despite the demands and risks of spaceflight, I can't imagine any other career so fulfilling and downright fun.

It is difficult to push back frontiers. It is difficult to build equipment that operates in the harsh environment of space. It is difficult to provide for the health and well-being of the people who live there. But that is precisely why I chose the astronaut profession. While I always strove for the best, I was prepared to deal with the worst.

9

THOMAS TOWNSEND

Visiting Scholar at the Centre on Public Management and Policy, University of Ottawa

Using the F-Word in Government

In public-sector circles, Thomas is known as a doer, as someone who gets things done. He's worked in the system: in Correctional Services Canada and the Canadian Mission to the European Union, and as Assistant Deputy Minister for the Policy Research Initiative. Currently, he is a visiting scholar at the Centre on Public Management and Policy at the University of Ottawa, and is also involved with the Executive M.B.A. Program at the University of Turku in Finland. Finding public-service executives willing to speak about failure is a complex undertaking, but luckily for this project, Thomas stepped up for the cause in a major way!

— Alex Benay

"Federal Sponsorship Scandal"
— CBC Online, October 26, 2006
"Adscam Reveals Deeper Government Mismanagement"
— *Maclean's*, July 19, 2006
"Residential Schools Called a Form of Genocide"
— *Globe and Mail*, February 17, 2012
"What Went Wrong in the Tainted Blood Disaster?"
— CBC Radio, March 17, 1992
"Why the Long-Gun Registry Doesn't Work — and Never Did"
— *National Post*, December 11, 2012

We have a failure rate. But part of it [is] because we're trying stuff nobody else would be willing to try. I think, probably it's the fear of failure that paralyzes people.[1]

Failure in government makes good headlines. Readers of this chapter are more likely to be familiar with examples of what has gone badly in the public sector than what has gone well. And failures are typically recounted as blameworthy stories of malfeasance, mismanagement, and incompetence. Failure in the public sector is portrayed as unacceptable under all circumstances.

It's noteworthy that when the focus is on something other than the public sector, attitudes about failure are different. We tolerate failures in our personal lives and in private enterprise. We comfort ourselves with expressions like "live and learn"; we tell our children "if at first you don't succeed ..."; we admire the vision and stick-to-it-iveness of inventors and entrepreneurs like Thomas Edison, who said:

> I have not failed 10,000 times. I have not failed once. I have succeeded in proving that those 10,000 ways will not work. When I have eliminated the ways that will not work, I will find the way that will work.[2]

Is it right to have a double standard about failure when it comes to the public sector? We leave that question open. What's indisputable, though, is that intolerance of failure in the public sector affects the ways in which public servants deal with failure. Failure is a taboo subject for public servants; it is virtually undiscussable. "Failure" is the "F-word" of the public service.

Some of us believe, based on long careers as federal public servants, that prevailing attitudes within the public service toward failure perpetuate a large and costly lost opportunity. Fear of failure — fear not only of failing but also of openly discussing failure — stifles creativity and innovation and inhibits learning. The public sector's chronic inability to learn from failure was documented by the Auditor General of Canada, who said in his 2016 report to Parliament:

> Our audits come across these same problems in different organizations time and time again. Even more concerning is that when we come back to audit the same area again, we often find that program results have not improved.... In just five years, with some 100 performance audits and special examinations behind me since I began my mandate, the results of some audits seem to be — in the immortal words of Yogi Berra — "déjà vu all over again".... [3]

This is the opposite of what is required in a rapidly changing world where governments need capacity to respond nimbly to emerging challenges. A dysfunctional relationship with failure is bad for the public service and for Canadians. It's time we got comfortable saying the F-word out loud in polite company.

If the problem described by the Auditor General is rooted in how the public service deals with failure, then what is the solution? That brings us to the purpose of this chapter, which is to propose an approach for building a healthier relationship with failure in the public service.

There is much to be gained for Canada and Canadians from building the capacity and willingness of public servants to value, and extract value from, failure. There is even a case to be made for encouraging public servants — within limits — to pursue actions that have a high likelihood of failure. (More on that later.)

Before going further, let us be clear that there are limits to altering the public service's relationship with failure. The Clerk of the Privy Council — the most senior federal public servant — recently said in an internal "town hall" meeting that "if you aren't failing you aren't trying hard enough." Although the message was welcome and well intended, some people in the room may have hesitated to take it seriously because they know that fear of failure in government is underpinned by some inalterable truths. Two are particularly important.

First, government departments are a public trust. They spend public money for public benefit under the leadership of elected ministers. Demands for stewardship combined with political pressures create a degree of public scrutiny, a level of accountability, and an acute sensitivity to public perception that will never abate (nor should they).

Second, everything the public sector does is intended, directly or indirectly, to contribute public well-being. Failure translates to a loss (real or perceived) of well-being. In extreme cases, such as tragedies like residential schools for First Nations youth or the tainted blood crisis, public-sector failure becomes a matter of life and death.

So while we need to be better at embracing failure in the public sector, we must do so with care and a hard-nosed sense of what is feasible.

How Do We Get to a Better Place?

The public service won't improve its relationship with failure as long as failure continues to be treated as a large, undifferentiated concept that looms like a formless grey sky. To that end, this chapter proceeds in three steps:

- First, identifying a few broad categories of failure and identifying one of them as the place to focus efforts on improving our relationship with failure.

- Second, isolating three general factors that warn that failure is likely.
- Third, proposing practical measures to build capacity for recognizing and acting upon the warning signs of failure.

Whose Fault Is It, Anyway?

A study[4] of Auditor General reports from 1988 to 2013 identified 614 cases of public-sector failure. It found cost overruns, mispayment of benefits, failure to meet objectives, misreporting of financial information, inefficiencies in service provision, breaches of policy and guidelines, spending on items that provided no value, and improper management.

It's fair to assume that only a minuscule proportion of the Canadian population was aware of more than one or two of the 614 failures. As anyone who has spent time inside the public service will appreciate, the vast majority of failures would be of no interest to headline writers. They are relatively insignificant in terms of harm caused, and are rarely the work of ill-intentioned culprits.

A study of failure in the private sector[5] published in the *Harvard Business Review* supports the view that for the vast majority of organizational failures there is little to be gained from seeking to "blame someone." The author, Amy Edmondson, concluded that assigning blame would have been useful in less than 5 percent of the cases she reviewed. What's most interesting is that the managers involved with the same set of failures saw things completely differently; they felt that establishing blame was useful in 90 percent of cases! (This suggests that the private sector may not have a healthier relationship than the public sector does with failure.)

Bad Failure; Good Failure

Edmondson's key insight — which is relevant to the Canadian federal public service — is that not all failures are created equal. She identifies a spectrum of failure from blameworthy to praiseworthy.

- **Failures that are preventable** are always bad; searching for blame may be relevant. These are failures where processes and procedures are standardized and repeatable. Programs delivering income benefits such as Old Age Security or Employment Insurance are examples that fit this description. In this type of environment the government department has a high degree of control over occurrence of failures, and causes of particular failures are relatively easy to isolate. Blame can, and arguably should, be established.

- **Failures that result from process complexity** tend to happen in times of changing conditions; it may be counterproductive to treat these as bad and blameworthy. Cumbersome government processes for interacting with the public — in an environment where customer-service innovations by companies such as Amazon and eBay have set the bar for public expectations — are an example of this type of failure. As well, processes can be rigid and impede staff from doing their job, so a choice between getting the job done and following the rules emerges. This type of failure may result from multiple causes over which no single individual, program, or organization has significant influence. Figuring out "who is to blame" may be more trouble than it is worth, and indeed may assign blame where it is not deserved.

- **Failures that occur when an organization is at the frontier of what it knows** are always good (when managed effectively). These failures may happen when government organizations decide to fundamentally rethink what they deliver, or how they deliver, or both. Experimentation and failure may be a necessary part of achieving successful transformation from an old way of doing business to a new and better one or launching a new service. The notion of "blame" is counterproductive here.

The focus of our discussion on building a healthier relationship with failure is in this last area — actions at an organization's knowledge frontier — where blame has little relevance and potential for learning from failure is high. The main challenges here will be to shift the conversation from "no failure allowed" to "failing safely," and to build capacity and procedures for early detection of failure that contribute to learning and succesful adaptation.

Failure at the Frontier – The Example of CESG

In the late 1990s, after three decades of significant increases in the cost of education, the federal government wanted to ensure that post-secondary education was accessible to all Canadians, especially students from low-income households. In 1998 it launched the Canadian Education Savings Grant (CESG), which created an incentive for families to save for their children's education. It was a "frontier" program because the federal government had never tried anything like it before. This is a program I was responsible for and can speak to from direct experience.

An evaluation of the program in 2003 revealed an unintended and undesirable outcome: higher-income households were benefitting from CESG at a much higher rate than lower-income households. In other words, a significant proportion of the program's resources were subsidizing savings by families who in all probability would have saved for their children's education in the absence of the program. In response, extra effort was put toward raising awareness of the program among lower-income households; in addition, resources were devoted to creating greater incentives for these households.

Notwithstanding the corrective efforts, an evaluation in 2015 still found that higher-income households were the primary beneficiaries. So, did the program fail? Although it no doubt opened access to post-secondary education for some lower-income households, the primary beneficiaries turned out to be families that would have managed without the support.

Could these unintended outcomes have been foreseen at the time when the policy — the decision to support access to post-secondary education, with an emphasis on low-income families — was first being discussed? Yes, to some extent they could have been forseen. Ways might have been found to direct more of the program's benefits to the most needy households if the approach to program design and implementation had allowed for greater operational experimentation.

Why, once it became apparent that the program was not unfolding according to plan, was it subjected only to tweaking rather than more radical redesign? Because in the absence of incentives to experiment with various redesign options "on the fly" — and the compounding absence of structures and procedures within which to conduct small-scale experiments — program managers understandably limited themselves to small, safe, path-dependent changes.

Anticipation — Yes; Trepidation — No

An advantage of choosing the knowledge frontier as the place to begin changing the conversation around failure is that it is *relatively* easy for people operating at the frontier to admit that their knowledge is imperfect, and that they may have to learn as they go. (This can be an incredibly difficult thing to say out loud in the public service!)

The "frontier atmosphere" — if we can call it that — makes it easier to create environments where failures can occur safely and where learning from failure can be captured and integrated into whatever may come next. The idea is to learn how to anticipate failure rather than fear it. This opens the way to early detection of failure followed by corrective action and rapid recovery. In short, you focus on identifying and dealing with problems before the public and the institution become exposed in a large way.

Failure in Operating Public Services — Warning Signs

Once programs and services are operational, failures can occur, but managers need to detect and correct them early. Are there warning signs of failure that program managers should be able to detect and act upon? A recent study[6] provides a relatively simple way to approach this question. It identifies three important sources of public-service failure:

- Ignorance
- Rigidity
- Neglect

Each of these terms is provocative, perhaps deliberately so, as the analysis suggests that any program manager on the alert for warning signs of failure should watch for one or some combination of these three things.

I. Failure Due to Ignorance That Change Is Necessary

The government's interaction with citizens — the manner in which it delivers services, provides benefits, disseminates information, and so forth — is where this type of failure is most likely to happen. Information systems too often don't provide sufficiently detailed insights into citizen experiences, or are not sensitive to changing expectations. As I found out in several operational jobs, reporting can often be about counts, not service.

In 1996 the Auditor General published the following findings about public-service delivery:

- many, if not most, departments had not developed formal service standards;
- the public service, with some exceptions, was not service oriented;
- new technology that might improve service delivery was underused;
- service considerations tended to be secondary to administrative ones; and
- many services did not collect and analyze data on complaints.

That was more than twenty years ago. How do things look now? In his 2016 report, the Auditor General lamented the continuing malaise affecting service delivery.

> Over the years our audit work has revealed the government's lack of focus on end-users, Canadians.... In an age of instant communications, Canadians expect quick results, while governments are often stuck using old, slow approaches that fail to meet expectations. The slow speed of government is an issue that we have reported on often, and we are

reporting on it again [now].... In a 2013 audit, we found that government online services were not focused on the needs of Canadians, and that accessing those services was complex and time-consuming.... Government departments and agencies need to look differently at their positions as service providers.... Government ... needs to be good at service delivery to remain relevant.[7]

It may seem difficult to accept that programs and services, in the face of so much evidence, may be ignorant of the need for change. (From the perspective of users, the need for change could not be more obvious!) But when the government delivers a service, it is typically a monopoly provider, and focus in the past was on administration. It is not inconceivable that a monopoly provider may be blind and deaf to evidence of the need to change, because absence of competition dampens an important incentive for paying attention to users. As a former senior Ontario government official has observed, "introduction of competition appears to be the most significant factor" driving quality improvement of public services.[8] The point here is not to advocate for more private provision of public services — that is an entirely separate discussion — but rather to observe that failure due to ignorance of the need for change is possible even when plenty of evidence of the need for change is available. That evidence is most clearly observed through direct, unfiltered contact with citizens receiving the service, and if that process is not in place and part of the regular management discussion, the danger of failure is present.

2. Failure Due to Rigidity

Consider public procurement — the set of rules and procedures whereby the federal government buys goods and services. I now do work for governments worldwide and must navigate the web of rules.

Businesses that sell to the government — especially small businesses like the one I am associated with — find the complexity of the procurement system profoundly vexing. Many small businesses simply forfeit the opportunity to sell to the federal government because the burden imposed by the procurement system is too onerous. As a consequence, those businesses lose access to potential revenue, while the government loses access to potentially innovative, and value-adding, goods and services.

Most goods and services are procured through competitive bidding processes, intended to ensure fairness for suppliers and generate "value for money" for the government. The intended outcomes — "fairness" and "value for money" — are laudable, but the processes ostensibly aimed at yielding those outcomes are antiquated and rigid. As an example (this is a story a heard directly from someone involved), a recent request for proposal (RFP) asked each bidder to provide twelve hard copies and eight soft (electronic) copies of the proposal. The link between the administrative requirement and the intended outcomes of fairness and value for money was not obvious. When the procurement officer was asked why the requirement for so many copies was necessary, he conceded that he could accept "eight hard copies and eight soft copies"!

The demand for a nonsensical number of copies is probably founded in some real moment of procurement history. There may have been a time (decades ago?) when it made sense to ask for eight or twelve copies. The fact that it *no longer makes sense* to do so, and that the notion of asking for multiple *soft* copies is ludicrous seems to have escaped the procurement authorities.

The reference to history — to the fact that some apparently useless requirement may once have made sense — points to a classic sign of rigidity. If pushed further to explain the requirement for so many copies, the procurement officer would inevitably have reverted to justification by historical precedent; he would have said, "We have always done it this way."

Every public servant reading this chapter will be familiar with the "we have always done it this way" defence. It is the most overused and least valid rationale for preserving the status quo in any area of government operations or policy. If ever you hear those words there is a high probability that you are face-to-face with a failure by rigidity.

Rules and procedures, in and of themselves, are neither good nor bad. They are instruments — means to an end. They are used in the public sector to ensure fairness and good stewardship, among other things. Failure by rigidity happens when rules and procedures remain in force even though they no longer serve their intended purpose, or have accumulated layers over time. What is interesting about rigidities is that they are best observed from the outside, while at the same time outsiders are often reluctant to provide an honest assessment to officals who may affect their well-being. In administration-intensive organizations I can safely assume that there are rigidities. We can run experiments using a reduced set of rule and procedures to observe what happens.

3. Failure Due to Neglect

The Great Lakes — storehouse of 20 percent of the world's fresh surface water — are victims of failure by neglect. Overrun by 185 invasive species — non-native aquatic plants and animals accidentally or deliberately introduced from outside — their ecosystem has suffered irreversible damage.

Canadian research published in 1991 concluded that discharges from ballast tanks of ocean-going ships at Great Lakes ports were an important source of invasive species. A solution was soon proposed: authorities in Canada and the United States should require ocean freighters to rinse their ballast tanks with seawater (salt water kills most of the biological residue in the tanks) before entering the St. Lawrence Seaway.

In both Canada and the United States, a combination of bureau-cratic inertia, inadequate legislation, spotty implementation of ballast water regulations, and shipping-industry lobbying allowed the problem to fester for fifteen years. During that time about twenty new invasive species attributable to ballast discharges were discovered in the Great Lakes. In 2002 the auditor general, through the commissioner of the Environment and Sustainable Development, described in stark terms the federal government's neglect in responding to challenges posed by invasive species in the Great Lakes and elsewhere:

> The federal government has not taken effective action to prevent the introduction of invasive species that threaten Canada's environment or to control or eradicate them.... Unfortunately, Environment Canada[9] has not succeeded since 1995 in co-ordinating a practical response to the problem. It has not obtained the key information that it needs to effectively oversee or co-ordinate the federal government's response. It has not identified the invasive species that threaten Canada's ecosystems, habitats, and species; their most important paths of entry; or the risks they pose to Canada's environment and economy. It has not put together a national action plan or secured agreement among federal departments on who will do what to respond to major risks. Nor has it ensured that it has the tools it needs to determine whether measures that have been taken are working. The Department needs to get on with this basic work. [10]

It wasn't until 2006, long after much irreversible damage had occurred, that Canada finally implemented an effective legal and regulatory regime to control the dumping of tainted ballast water into the Great Lakes. The U.S. government was similarly slow to respond.[11] Neglect can happen by virtue of overly slow processes or by focus being redirected to other priorities. Again, it is best viewed from outside the organization.

Neglect can occur at smaller levels. Staff are good at keeping it together until they can't. Budget reduction exercises where there is not a corresponding effort to reduce workload create a hidden form of neglect. Large flat organizations can make it easy to take your eye off the ball, especially if certain areas are constantly demanding attention. Those quiet backwaters that all of us in management appreciate as they don't require our attention are potential sources of failure.

Talking About Failure, Safely

We have suggested that programs operating at the knowledge frontier are where effort should be focused on building a healthier relationship with failure. The frontier is where, as Edmondson has observed, there is "a space of psychological safety" within which it is okay for public servants to talk openly about failure.

We have identified three categories where public servants should be alert to warning signs of failure in operating programs and services: (1) ignorance; (2) rigidity; and (3) neglect.

The combination of these ideas leads us to conclude that a practical way forward is to identify readily implementable ways of building capacity to talk about, plan for, anticipate, and learn from failure in areas of public-sector activity that are at the knowledge frontier.

We are not proposing broader strategies for "culture change" related to failure, although some of what we propose may turn out to be the first steps toward culture change. Any culture change is a slow process requiring sustained leadership over a long period of time. We choose instead to focus on discrete actions that can be undertaken immediately, do not necessarily require senior-level leadership, and have the potential to deliver benefits in a relatively short time.

Recommendation 1: Encourage Discussion About Failure Before Failures Have Occurred

You are a program manager in the preliminary stages of designing a program. You bring together a group of people who will be involved in the program. You ask them to imagine that the program is already running and has failed spectacularly, and that the purpose of the meeting is to investigate the failure.

Participants take a few minutes on their own to write out reasons for the failure. The analyses are then shared, and proposed reasons for failure are categorized according to the degree of control the program team would be assumed to have had over them. The findings are fed back into the program's design. The exercise of anticipating failure *before* the launch of the program helps the program team identify weaknesses in the design.

This technique, called a "pre-mortem," was developed by Gary Klein, a behavioural psychologist with many years' experience working with first responders.[12] Research has shown that engagement in this type of thinking — sometimes referred to as "anticipatory hindsight" — promotes richer discussion by stimulating people to hypothesize significantly more reasons for failure than they would have done without using the technique.[13] Furthermore, it provides a safe way to discuss failure because the "failed program" has yet to launch. Failure scenarios are realistic but hypothetical; no one can be blamed for anything; no one risks suffering loss of reputation or damage to their career.

It is difficult to know if pre-mortem exercises were undertaken in the early stages of designing the federal government's failed programs whether failure could be averted in every case. But we can easily imagine how pre-mortem thinking, executed properly, would have made an important positive contribution to many projects.

Recommendation 2: Encourage "Safe to Fail" Experimentation

The increasing complexity of government operations means that there are a growing number of situations where the evidence does not give managers a clear indication of *what* a program should be delivering or *how* it should deliver in order to contribute to outcomes. This is characteristic of life at the knowledge frontier. It is an argument for running experiments to see what works and, more importantly, what does *not* work.

Public servants sometimes feel that they have engaged in experimentation through the pilot projects they run. To be clear, we regard a pilot project as being different from an experiment. A pilot project tests a relatively advanced program design; its purpose is to prove that the project concept works. Outright failure would be moderately surprising (at the very least) once things have advanced to the pilot-project stage. A pilot is usually expected to reveal certain deficiencies that fall well short of failure, or alternatively it may bring to light opportunities for a higher level of performance that can be addressed prior to full-scale implementation.

An experiment, on the other hand, is a probe to test an early-stage idea. There is no necessary expectation of success or failure: either outcome will be beneficial if it provides an opportunity for learning. Dave Snowden[14] has pioneered the use of safe to fail experiments and developed an approach that allows for measuring experiments to promote learning and ensure that they are curtailed if unsuccessful and scaled if successful.

Given the taboo in the public service regarding failure, and the significant probability that experiments will fail, it is important that experiments be conducted in a safe psychological space, where public servants feel comfortable allowing failure to happen. Many departments have opened "design labs" based on principles of design-thinking and behavioural economics. The labs allow staff to explore innovative ideas in an environment removed from the operational mainstream. These labs would be an ideal place to assign staff for the purpose of designing, launching, and monitoring experiments.

The purpose of an experiment might be, for example, to test alternatives for an entirely new program-delivery mechanism. The experiment might push the mechanism to the point of failure, generating valuable evidence about both its robustness (resistance to failure) and resilience (capacity to recover from failure). And experiments would be designed as safe to fail — that is, maximizing potential for operational learning while minimizing the risk of harm to the public and the department.

Again, we can't help but think that it would have been both feasible, and extremely useful, to have run small-scale experiments *aimed deliberately at producing failures* in prototypes of the programs and services before they where implemented broadley. In order to minimize conflicts of interest in the experimentation process, and minimize stigma or ill-feeling that might be associated with a deliberate pursuit of failure, the experiment team could have been assigned to a lab with its own management group and set of reporting relationships.

Recommendation 3: Reduce Insularity

When managers get caught up in day-to-day imperatives of their own organization, they risk isolation from the external world they are supposed to be serving. Insularity is likely to be an important accompaniment to each of the three reasons for failure highlighted earlier: ignorance of the need for change, rigidity, and neglect.

Insularity obstructs a healthy attitude to failure. It supports the management pathology of "groupthink"[15] — emphasis on reaching consensus and avoiding conflict through suppression of alternative points of view and isolation of the group from outside influences. A logical consequence of groupthink is an "illusion of invulnerability": everything is fine.[16]

There are some practical approaches to combat insularity for the sake of preserving openness to the idea and value of failure. Our suggestions follow the practice of "vigilant problem-solving"[17] where one deliberately seeks ideas and advice that "bring to a discussion diverse knowledge and opinions; and where the leader solicits dissenting views, critically examines costs and risks of the preferred choice, and is willing to revise an initial view in response to the deliberations."[18]

Peer review can be an effective and low-cost instrument for defeating insularity. Informal peer review — of, for example, early-stage proposals of a new program or policy — need not be difficult to establish and operate.

Within departments it would be a relatively simple matter to establish cross-divisional teams to do peer review for each other on an informal basis. In departments with regional offices, staff in different regions from the same program could be enlisted as peer reviewers. To encourage a greater diversity of opinions, it would be worth considering use of peer-review teams from other departments. Different departments face similar challenges in program design, service delivery, and administrative functions such as procurement, so there is much to be gained from cross-departmental peer review.

The United Kingdom has Policy Profession, an informal cross-government network of public servants who work in, or are involved with, the formation of policy for government. Its goal is to build capability across the public service for:

- development and use of a sound evidence base;
- understanding and management of the political context; and
- planning and clarity as to how the policy will be delivered.

The Ontario government has a similar network of policy professionals. Such groups offer connections to peers who can talk constructively and without bias about the strengths and weaknesses of proposed initiatives, and whose views can help break through the insularity that may contribute to failure.

Recommendation 4: Integrate Capacity Development for Learning from Failure into Management Training

Currently there are no training programs available to Canadian federal public servants oriented toward building a healthier relationship with failure, or featuring skills development related to learning from failure.

Integrating courses of this type into the curriculum of the Canada School of Public Service would serve a dual purpose. First, and obviously, it would help build acceptance within the public service of the idea of valuing failure, while also strengthening capacity to learn from failure. Second, it would send a signal that the public service is serious about looking at failure in new and constructive ways.

Conclusion

Now is the time for Canada's federal public service to take a healthier, more constructive approach to failure. Nowhere is this more evident than in the prime minister's current mandate letter to the manager of the public service, the president of the Treasury Board, instructing him to encourage ministerial colleagues to "[devote] a fixed percentage of program funds to experimenting with new approaches to existing problems."

This directive has far-reaching implications. Experimentation requires openness to failure. Discovering "new approaches to existing problems" will inevitably mean (as Thomas Edison would have put it) identifying "10,000 ways that don't work" before identifying the one way that does work and yields a major breakthrough in efficiency and effectiveness.

The prime minister's bold call for experimentation lands public servants in unfamiliar territory. They are being asked to become, within reasonable limits, comfortable with failure — comfortable with talking about it, valuing it, admitting that it has occurred, learning from it. Management philosophy needs to evolve from a single-minded emphasis on robustness (preventing all occurrence of failure) to one of resilience (recognizing and recovering from failure).

It is unrealistic to expect that the evolution will occur overnight. Old habits die hard, and the instinct to flinch at the thought of failure is an old habit indeed. With this in mind, the recommendations we have made in this paper balance our public-servant-minds' understanding of what is possible over the short term with a more aggressive desire to see meaningful change happen sooner rather than later.

A head of corporate communications at a large public organization once shared his belief that major organizational change sometimes begins innocuously at the level of vocabulary. "People start using words that suggest change has happened, even if it hasn't happened yet. That's the first step," he said.

There is truth in that. And so let us look forward to a day in the not very distant future when it will be okay to say the F-word in polite company in the Canadian federal public service.

Notes

1. Anonymous high-performing United States government senior executive quoted in Kelman, et al. (2014).
2. Edison is widely quoted as having said words to this effect; several variations of this quotation can be found. We have not been able to find the original source. This version is taken from Furr (2011).
3. "Message from the Auditor General of Canada," *Report of the Auditor General of Canada to Parliament,* 2016.
4. *Federal Government Failure in Canada, 2013 Edition: A Review of the Auditor General's Reports, 1988–2013.*
5. Edmondson (2011).
6. Van de Walle (2016).
7. "Message from the Auditor General of Canada," 2016.
8. Dean (2011).
9. The name of this federal department has since been changed to Environment and Climate Change Canada.
10. Office of the Auditor General of Canada (2002).
11. For a detailed account of neglect by the Canadian and U.S. governments in responding to invasive species in the Great Lakes, see Alexander (2009).
12. Klein (2007).
13. Mitchell, et al. (1989).
14. Snowden (2007).
15. Janis (1982) is a seminal work on the subject.
16. Janis and Mann (1977).
17. Janis (1989).
18. Kelman, et al. (2014).

References

Alexander, Jeff. *Pandora's Locks: The Opening of the Great Lakes — St. Lawrence Seaway*. East Lansing: Michigan State University Press, 2011.

Dean, Tony. "Is Public Service Delivery Obsolete?" *Literary Review of Canada* 19, no. 7 (2011). http://reviewcanada.ca/magazine/2011/09/is-public-service-delivery-obsolete/.

Edmondson, Amy C. "Strategies for Learning from Failure," *Harvard Business Review* 89, no. 4 (2011): 48–55.

Eggers, William D. "Faster Government: Rethinking the Risk Equation." *Governing*, August 30, 2016. www.governing.com/columns/smart-mgmt/col-faster-government-power-failing-well.html.

Employment and Social Development Canada. *Canada Education Savings Program: Summative Evaluation Report, Final Report*. Ottawa: Government of Canada, 2015.

Ferguson, Michael. "Message from the Auditor General of Canada." *Office of the Auditor General of Canada*, 2016. www.oag-bvg.gc.ca/internet/English/parl_oag_201611_00_e_41829.html.

Furr, Nathan. "How Failure Taught Edison to Repeatedly Innovate." *Forbes*, June 9, 2011. www.forbes.com/sites/nathanfurr/2011/06/09/how-failure-taught-edison-to-repeatedly-innovate.

Gavett, Gretchen. "When We Learn from Failure (and When We Don't)," *Harvard Business Review*, May 28, 2014. https://hbr.org/2014/05/when-we-learn-from-failure-and-when-we-dont.

Holgeid, Kjetil., and Mark Thompson. "A Reflection on Why Large Public Projects Fail." In *The Governance of Large-Scale Projects: Linking Citizens and the State*, eds. Andrea Römmele and Henrik Schober, 219–44. Baden-Baden, Germany: Nomos Verlag, 2013.

Human Resources Development Canada. *Formative Evaluation of the Canada Education Savings Grant Program: Final Report*. Ottawa: Government of Canada, 2003.

Janis, Irving L. *Groupthink: Psychological Studies of Policy Decisions and Fiascoes*. 2nd ed. New York: Houghton Mifflin, 1982.

Janis, Irving L., and Leon Mann. *Decision Making: A Psychological Analysis of Conflict, Choice, and Commitment*. New York: Free Press, 1977.

Kelman, Steven, Ronald Sanders, Gayatri Pandit, and Sarah Taylor. "'Tell It Like It Is': Groupthink, Decisiveness, and Decision-Making Among U.S. Federal Subcabinet Executives," HKS Faculty Research Working Paper Series. August 2014.

King, Anthony, and Ivor Crewe. *The Blunders of Our Governments*. London: Oneworld Publications, 2014.

Klein, Gary. "Performing a Project Premortem," *Harvard Business Review* 85, no. 9 (2007): 18–19.

Lammam, Charles, Hugh MacIntyre, Jason Clemens, Milagros Palacios, and Niels Veldhuis. *Federal Government Failure in Canada, 2013 Edition: A Review of the Auditor General's Reports, 1988–2013*. Vancouver: Fraser Institute, 2013.

Light, Paul C. "A Cascade of Failures: Why Government Fails, and How to Stop It." *Brookings Institution*, July 14, 2014.

Mitchell, Deborah J., J. Edward Russo, and Nancy Pennington. "Back to the Future: Temporal Perspective in the Explanation of Events." *Journal of Behavioral Decision Making* 2, no. 1 (1989): 25–38.

Office of the Auditor General of Canada. *2002 October Report of the Commissioner of the Environment and Sustainable Development: Chapter 4 — Invasive Species*. Ottawa: Government of Canada, 2002.

Shergold, Peter. *Learning from Failure: Why Large Government Policy Initiatives Have Gone So Badly Wrong in the Past and How the Chances of Success in the Future Can Be Improved*. Canberra: Commonwealth of Australia, 2015.

Snowden, Dave. "Safe-Fail Probes." *Cognitive Edge*, November 17, 2007. http://cognitive-edge.com/blog/safe-fail-probes/.

Van de Walle, Steven. "When Public Services Fail: A Research Agenda on Public Service Failure." *Journal of Service Management* 27, no. 5 (2016): 831–46.

Wajzer, Chris, Oliver Ilott, William Lord, and Emma Norris. *Failing Well: Insights on Dealing with Failure and Turnaround from Four Critical Areas of Public Service Delivery*. London: Institute for Government, 2016.

10

ERICA WIEBE
2016 Olympic Gold Medalist, Women's 75 kg Freestyle Wrestling

How Failure Led Me to Olympic Gold

I met Erica last year, barely a week after her gold medal win at the Rio Olympics. What struck me about her was her humility. This was someone who had just won a gold medal, for God's sake, and her aura was one of humility and appreciation. We had just acquired some of her training equipment for the wearable technologies exhibition we were developing for the launch of the Canada Science and Technology Museum, and she was stopping by to do a media event to promote our new partnership. She made her way through the staff and the dozens of youth who were in the room and quickly dispensed with the official part of the morning, making her way back toward the kids. She spent most of her time answering their questions, taking pictures, and signing autographs. During that brief time with our museum youth, she talked about her efforts, her mistakes, and the dedication required to win an Olympic gold medal. There were too many amazing life lessons to enumerate in such a short span of time!

That's why I asked Erica to participate in this book project — because I saw her humility and heard about her journey (which included quite a few fails, by the way). I could see that she was the perfect ambassador to speak to her own failures — and to influence other young girls to take up sports in our amazing country. Canadian sport is about more than hockey; it's about athletes like Erica who fail, get back up, and try again, day in and day out. She is a natural fit for this book.

— Alex Benay

In sport, failure is all but guaranteed. You have a better shot at getting hit by lightning than going to the Olympic Games. You want to stand on top of the podium at the Olympic Games? Out of the 11,237 athletes who competed at the 2016 Summer Olympic Games in Rio de Janeiro, 306 won gold. That means after beating all the odds and qualifying for the Games, if you consider winning as the barometer of success, there is still a 97 percent probability of failure.

Fortunately, winning or losing does not define failure. When I look back upon my experiences in sport and life, it is my failures that have helped to shape who I am and where I am going. Because I made mistakes, failed often, and accepted that failure was never fatal or final and because I knew that with each failure, I would learn, evolve, and grow, I was able to stand on top of the podium at the Olympic Games.

When I started wrestling, women's wrestling was not an Olympic sport, so I did not grow up with the ambition of standing on top of the Olympic podium in the same way many kids today confidently claim. In grade nine there was a sign posted outside my gymnasium doors; it said, "Co-Ed Wrestling Practice, Monday." I quickly learned that in the sport of wrestling, there are no cuts to make; whoever survives makes the team. It is the same in most high-school and university-level programs in Canada. We need bodies to train, and the ones whose will is strong enough to keep getting up and coming back are those who have the best ability to succeed.

That invitation to attend my first wrestling practice presented an opportunity (in my eyes) to wear spandex and wrestle with boys. I was an active kid who was always open to trying new things. I liked the spandex and immediately connected with a sport that required so much mental, physical, and emotional strength. That connection turned to passion and quickly translated into success. Two years into the sport, I won a spot on the U18 Cadet National Team. At a team training camp in the summer I was having fun on and off the mats, but at that age I did not yet have the discipline to be a true champion. I broke curfew one night and two days into the camp was sent home. I was devastated.

Alone on the bus for the four-hour-long ride back to Edmonton to get on a plane going home, I stared out the window and was deeply disappointed. I had failed my coaches, my teammates, and myself. My own actions had resulted in my not being able to do what I loved most.

I resolved to be better. To never let myself get in the way of what I love doing most. That was a powerful lesson that I would be reminded of years later, but in much more dramatic circumstances.

o o o

I moved back to Alberta in the fall of 2007. Bright-eyed and ready to make an impact at the University of Calgary, I returned to Alberta because it had the best women's wrestling program in the country, and arguably the world. I wanted to be training with the best coaches, facilities, and training partners. That whole first year in Calgary I did not score a single point in practice. I was training with teammates older, stronger, faster, and more experienced than I was — and I would fail. Every day.

I do not know what kept me going through that first long Calgary winter. Eventually March rolled around and I had the opportunity to wrestle at the Junior Canadian Nationals. I would finally be competing against women who were at my level of experience and my own age. I knew that not a single one of them had been taken down as many times as I had that past year, but importantly, I knew that none of them had gotten up as many times, either. I had confidence that I had done everything I could, so I just went out there and wrestled. I won.

Wrestling teaches you resilience unlike any other sport. In an incredibly intimate way, you are engaged in the most primal form of physical combat, in which the requirement to struggle, to compete, to fight, to triumph, is systematically built into the sport. As I would come to learn, wrestling teaches not just physical resilience, but also another kind that only you have control over.

o o o

In 2013 I won the Senior Canadian National Championships, but had to secure my spot on the 2013 World Team by winning a wrestle-off weeks later in Guelph, Ontario. The wrestle-off would be against my long-time teammate and the 2012 Olympian, a wrestler I had never beaten. That experience marked the first time that I made a decision not only to commit to the process physically but also to prepare myself mentally. This involved spending many hours visualizing myself being successful. Every night before bed I would visualize myself succeeding in various situations, and visualize myself having trouble in situations and rising above the challenge. I would visualize her coming at me and performing an offensive move on me, but I would take over that move and take her down.

Finally, the weekend of the wrestle-off arrived and I was ready to compete. As per my normal pre-competition routine, I set out my equipment the night before and was looking forward to a good night's sleep. I was excited yet relaxed. My match was at 9:00 a.m. I woke up the following morning to the sound of my hotel phone ringing, and I thought, "That's weird!" I wiped the sleep out of my eyes, rolled over, picked up the phone, and looked at the clock. It said 8:43. I put the receiver to my ear and the national team coach shouted, "Where the %&*$ are you?!" "Oh my god!" I thought, and threw everything into my bag and rushed over.

Luckily, we were only a two-minute walk from the event venue. I ran over to the gym, put on my wrestling boots, and had two minutes to spare before I was supposed to walk onto the mat, hoping to make my first Canadian senior national team. My adrenaline was surging in that moment, but I had already mentally wrestled that match a thousand times. I already knew what the outcome was going to be. I had never been so confident, never felt so comfortable — and I had never won a match so easily, so seamlessly. There was something about using visualization, not only of successful wrestling moves, but also of mentally seeing myself rising above the challenge, that gave me confidence.

Later that year I began making my mark on the international scene. I beat the 2012 Olympic champion in August and I put myself on the map as a major threat in my weight class. The following year I continued wrestling tough and beating international opponents. I won the Commonwealth Games, I won the University World Championships, and I had an unbeaten streak of ten straight tournaments. I had beaten everybody in the world in my weight class.

My performance results told the story of an athlete of the top of her game, but I read them differently. I had begun to focus too much on winning and was defining myself by each outcome on the mat. I had a perfect record, but deep inside I knew I was far from perfect. I felt I was living a lie; I was vulnerable and raw but I could not admit it to anyone. I was a winner. I was having tremendous success. How could I have so much doubt? I began to fear the failure that would break my perfect record, and I avoided facing that failure at all costs.

The 2015 Canadian Nationals took place in March of that year. The results of that one-day competition would dictate the Pan American Games team and the 2015 World Team, and be the first qualifier for the Canadian Olympic Team trials.

I had been having so much success internationally, but I did not feel complete. There was so much focus on winning, so much expectation of success, and I needed these past successes to give me confidence. However, I failed to look within myself and understand who I was and what were the intangibles leading to this success. I was working hard — no one works harder than I do — but something wasn't right. When I walked into the Nationals I had never felt so certain of my own defeat. And so, that day I lost. I was ranked top three in the world, but I would not be representing Canada at the 2015 Pan American Games, or at the 2015 World Championships. I was not going to be number one in Canada that year.

I walked off the mats and out of the gym that day, passed the shocked faces of my coaches, passed my parents, and passed the Canadian wrestling community, and I felt better than I had in months. I had lost. I had failed. But that failure released so much burden from my shoulders. I realized that I had been identifying myself as a winner, and as successful solely because of the medals I had won and nothing else. That did not feel right to me. I realized that the people who mattered most to me, and who cared about me the most, loved me and cared for me whether or not I was a winner. It was really important for me to realize that nothing was going to change about who I was, whether or not I won that wrestling tournament. My wrestling victories had directed my focus

Erica Wiebe works for the takedown against Vasilisa Marzaliuk of Belarus in the semifinal at the 2016 Olympic Games.

onto the medals and the podiums, and away from who I was as a person, from the motivations that fuelled me, and ultimately, what success really meant to me. Through reflecting on that loss and the emotional journey I had been on in the days leading to it and afterwards, I realized that going forward I would need to change my definition of success.

I shifted away from tangible outcomes and toward the pursuit of something deeper. I was no longer going to be defined by the outcome of one day, by one win or one loss. I was going to be defined by who I was and the person I had become through this journey. I examined the reason I had started down this crazy path in the first place. I was never motivated by winning or losing; I simply loved the sport. I set upon preparing for the Canadian Olympic Team Trials with a new definition of success. I returned to training, to pushing myself beyond my limits each day, but I entered the competition with a different intention. I could not guarantee I was going to win, but I could guarantee I was going to be successful.

◦ ◦ ◦

In December 2015, two days after winning the Canadian Olympic Team Trials, along with the Women's Olympic Wrestling Team I attended the Olympic Excellence Series with a hundred or so other Canadian Rio-hopefuls. We arrived in Toronto and were immediately brought into the opening session. All athletes were given a small piece of paper and asked to write a note to themselves that would be sealed and brought to Rio. I wrote:

> *I promise to step onto the mat in Rio and be my best self. I will leave no stone unturned and I will push myself beyond what I thought was possible. When I walk on the mats, I will already be a champion.*

That same weekend, we listened to Mike Babcock, Clara Hughes, Deidra Dionne, and a handful of Canadian sports icons share their Olympic experiences and provide insight and guidance for our journey over the next eight months. It was what Colonel Chris Hadfield said that impacted me the most. He spoke about how as astronauts they spend months, and even years, training each day for the worst possible scenario. In a spacecraft waiting to be launched out into space, strapped onto a rocket, every second you must combat a situation in which your decision means life or death. So they train on how to deal with that awareness and how to recover from failures. They consider every possible failure, so that what remains is the best possible chance of success.

This resonated with me. When I fail in my work (which I do often, and at times, spectacularly), it is not life-or-death, but there is always something on the line. Throughout my life, visualizing failure has been a part of me. As a kid, I would imagine my house catching fire, and I would go through the various ways I could save myself and my dog and cat. (I guess my family members were going to be left to fend for themselves!) But, through those very early visualizations, I began working through possible outcomes. Later on, as an athlete, it was my ability to address failure and risk, and to persevere, that gave me the strength to understand that failure is never final.

Failures come in many shapes and sizes: small ones, big ones, seemingly inconsequential ones that snowball over time. But sometimes they can come at a breaking point. My breaking point arrived seven days before I was supposed to fly to the 2016 Olympic Games. I failed myself again on July 26, 2016. On that day I found myself crouched on a bathroom floor, soaked in sweat, blood pouring down my face, sobbing.

Two hours earlier I had given up. Mentally, physically, emotionally, I was exhausted. I was standing in front of the mirror, braiding my hair, but I could not look myself in the eyes. I was one of Canada's great medal hopefuls in freestyle wrestling. I was supposed to continue the strong legacy of women's wrestling in Canada, to be the next Olympic hero in a sport that, despite being under the radar within the Canadian sporting context, has had one of the strongest Olympic performance histories of any summer sport. I was supposed to be unbeatable. I was supposed to be unbreakable. But I was broken.

I could not look at myself in the mirror because, although this was my dream, all I wanted to do was quit. If I could not score a point in practice (which I hadn't in ages), how was I supposed to score at the Olympic Games, where I would be competing in three weeks' time? I could barely lift my arms, and my heart was heavy. Throughout my career, one of the biggest gifts I have had is my spirit. I am the one who is always happy, always looking on the bright side, no matter what the circumstances. But for the first time in my career my spirit was crushed.

I finished braiding my hair and walked into the wrestling room to practise, hoping that just showing up was enough that day. I had nothing left in me to give. Preparations had been so gruelling; the demands on my body had finally taken their toll. Inside, I was begging our coaches to finish the endless weeks of training. They had reached a new level of insanity on the mats, and it continued today. Once again, we were going to be doing a very heavy-loading phase of wrestling, in which the volume and intensity of the wrestling reached their peak in the four years of Olympic preparations. Every new sparring partner was fresh and eager to score, and I had reached my limit.

Fighting back tears, I entered the final ten minutes of practice and I began wrestling with another fresh body. I stayed in my wrestling stance, knees bent, forcing my feet to move and block each one of my opponent's attacks. Keeping my hands moving, I felt that I was drowning and clawing myself to the surface, but instead I was clawing at my opponent, trying to grab a hold on a sweat-drenched body part and just hold on. Hold on to whatever I could.

My sparring partner changed levels, bending down and shooting forward. By instinct — only because I was too exhausted to think — I met my partner's level and blocked the incoming attack with my head. *Boom!* Our heads collided in an instant, searing flash. I managed to stay on my feet, but my face began to feel warm and wet, and I put my hands up to feel the blood pouring out from my brow. I turned and walked out of the wrestling room.

I did not speak to anyone. I just left.

It was just over one week from the day we were to head off to Rio, and I was lying on the table having my face stitched back together. It took just six stitches to close the cut along my upper brow, but I got up from that table and it felt like much more than that had been done.

I had failed so many times before, and although I could quantify the feeling I had on that day as "rock bottom," the one thing that I had learned in this sport was the ability to get back up and keep moving forward.

I lay there and reflected on what had gotten me to this point and what was at the root of my fears. Having qualified for the Olympics and knowing I was going to Rio, the struggle of who I was re-emerged. It was a struggle against others' expectations. Although I had been training in the same environment for the previous eight years, suddenly I felt overwhelmed by the expectations. I felt pressured to be perfect in training because I was going to the Olympics, and I was a medal hopeful, and I should be this amazing, perfect Olympic wrestler.

I could feel this pressure mounting and thought back to the Canadian Nationals in 2015, and how I had felt so much internal pressure to be successful. My winning streak had been perfect, but I remembered how I felt, and I reminded myself that I was *not* perfect. I made a choice that I would not be defined by success or failure. I had been through too much to be defined by others' measures of success. The only failure that truly mattered was failing to be my best self on each day of the journey. It was not easy, but I focused on all

Erica Wiebe, overwhelmed with emotions, looks down at her gold medal for the first time at the 2016 Summer Olympics.

the aspects of what that success would look like, and on only the things I could control. I decided that the only person I could fail was myself, and I was not going to let that person down.

o o o

I have taken you on a brief journey of the impactful moments in my wrestling career. I chose not to write specifically about the moment of winning the gold medal, because although it is a defining moment in my life, it does not truly define who I am. However, I will always remember the feeling I had on my day of competition at the Olympic Games because I woke up that morning and I already felt like a champion.

That morning I felt free from the possibility of failure. I had reframed my concept of success. Success was not tied to outcome but was woven into the process of preparation. I would not be defined by the outcome that day. I would be defined by who I had become. I had experienced failure many times over the years. I had spent the last three weeks of training at the edge of breaking, but I had done everything that had been demanded of me. In the final days before the Olympic Games, as part of my preparation I met one last time with my sports psychologist and confessed for the first time, aloud, that I did not think I could do it. It was a fear I had been feeling for a while, but it felt different to say it out loud, it felt so honest. He said to me that after everything I had been through, all of the struggle and the heavy work, *now* was going to be the really uncomfortable part. I had struggled and I had failed, but every time I had refused to give up and kept pushing harder. He said, "That voice in you that says '*push*' must now say: '*I am unbeatable.*'"

In some ways, it was easy to push each day. That was what I had done every day of the journey; it is my desire and capacity that differentiates me from the rest. However, believing in myself and what I am capable of was always a struggle. It is uncomfortable as a Canadian to be brash and unashamed of your talents and skills.

I am sure many people think that the purpose of a coach is to show you how to win the competition. If you do not win first place, the coach has also failed. However, the coach's job is to help you create a successful journey for yourself, so that you meet your own personal expectations of success. My coach, Paul Ragusa, has been quoted saying, "We don't chase medals, we chase performances." It may end in standing on top of the podium at the Olympic Games, but that outcome is more often than not outside of our control. Winning and losing do not translate into success and failure. When I tell people that my coach was very adamant that he did not care if I won a single match at the Olympics, they are surprised, even shocked. He just wanted to make sure that on that day, I wrestled my best. Early on he said, "I'm not going to ask you to do anything that I do not believe that you can do." Many times along the journey I was very much exhausted mentally and physically, but I knew that all I had to do was get through that day, and that was going to be enough. He talked a lot about grit, about just bearing down and doing what it takes on that day. In each wrestling match, the mantra was: one takedown at a time.

∘ ∘ ∘

We all fail in our lives. Some fail more than others. Denis Waitley says that failure is something we can avoid by saying nothing, doing nothing, and being nothing. I have seen many people who were afraid to fail. I learned many times that it takes courage to accept failure and evolve. It is uncomfortable, but as an athlete, I think that my willingness to be "uncomfortable" on a regular basis has been my defining feature.

There is one other story that I want to share in this chapter that is very important to me — important because it touches on failure in a different way. Not all failures have a silver lining.

I was in London for the Olympics in 2012 as a training partner with the Canadian wrestling team, and we had a taper camp just outside of the city. We would not be competing until the second week of the Olympics, but the athletes were in Great Britain to acclimatize and taper down the intensity of our training with the intention of peaking perfectly for that one day of competition. Our taper camp was emotionally tense: it was meant to physically prepare the wrestling bodies for optimal performance. We were given one day during that ten-day taper to go into London. It was a free day to do whatever we wanted — check out an event, or go see the venue. Two of my teammates and I decided that we were going to see the wrestling location and familiarize ourselves with the Olympic layout. Our day off coincided with the women's weightlifting events — specifically, the 63 kilo weightlifting class. We managed to sneak into the weightlifting venue due to a lucky synchronicity in accreditations. The weightlifting code was WL and the wrestling code was WR, and I guess the security guy thought he saw WL. He let us walk through and sit down in the athletes' section. So we sat down and began watching the weightlifting event just as they were finishing the last flight.

There were five or six lifters left to finish their sixth and final lift when the Canadian came onto the platform. Her name was Christine Girard and she was in fourth place. Weightlifting is such an exciting sport to watch as the crowd hushes to watch the display of instantaneous strength and power. She was doing the clean and jerk. She had 133 kilos on the bar and she cleaned it (got it up), jerked it over her head, held it long enough, and then dropped the bar. As a fellow athlete, I felt a deep connection with what I read on her face.

She had completed her athletic performance at the *Olympic Games*, and she felt so happy to be in *that* moment. And then she looked over to her coach. He was going absolutely insane — she realized at that instant that her lift was enough to bump her into the bronze medal position. She had just won Canada's first Olympic medal in women's weightlifting. I will never forget the first time I witnessed a fellow Canadian athlete win an Olympic medal in person. As an elite athlete our dream is that coveted Olympic medal. The iconic image of standing on the top of the Olympic podium is the representation of sporting excellence that is etched into our minds from a young age; and on that day Christine Girard *was* that icon, and made history.

That is not the end of this story. Four-and-a-half years later, after the London Olympics, after all of the retests of the 2012 athletes' samples, Christine Girard's bronze medal will turn to gold.* Because of two other people's actions and subsequent failures, Christine is going to get an Olympic gold medal.

This leads us to a form of failure that cannot be ignored when discussing the world of sport. A form of failure that goes beyond the individual experience and affects the true spirit of sport. Athletes' failures to pass drug tests have given us some of the definitive socio-cultural sporting moments of our times. Not even Usain Bolt's legacy is untainted by such failures. Cheating exists on the playground, and it exists at the highest pinnacle of sport; some people will do anything to win, at any level of play. The failure to play by the rules governing amateur sport across the globe robs athletes who do not cheat of their moment. Christine Girard will never get her gold medal moment standing on top of the Olympic podium. Failed drug tests in high performance sport represent a failure of the system at the nexus of sport and business and violates the true spirit of sport.

o o o

In some ways, failure is part of the Canadian consciousness. The Canadian experience at the Olympic Games used to be enveloped in the idea of *almost* — of coming up just short (I have heard it called a

---

* In July 2016, the IWF reported that the reanalysis of samples uncovered prohibited anabolic agents in both the gold and silver medalist, and on Wednesday, April 5, 2017, it was confirmed that Girard would become the first ever female Olympic weightlifting champion from Canada.

"Maple Medal"). We always seem to have the stories of just missing the podium. In 2004, however, when Vancouver won the bid for the 2010 Winter Olympic Games, a small group of individuals set out to change this paradigm with the bold vision that came to be known as Own the Podium. Canada went on to win the most gold medals at that Olympics. So, when I moved across the country to join the high-performance training centre for women's wrestling in Calgary, I was able to benefit from this cultural shift toward excellence in Canadian sport. I am a recipient of Own the Podium funding because of my sport's track record of Olympic success.

Beneath the surface of the Own the Podium program, which Canada has adopted as its own Olympic mantra, is a journey that is undeniably shaped by failure. It was the failure to win a single gold medal on home soil in both 1976 and 1988 that pushed Canadian sports leaders to seek a different path for Canadian high-performance sports as we forged our way toward the 2010 Winter Olympic Games. We achieved that goal in 2010 and have continued through subsequent Olympic quads. The new vision for high-performance sports in Canada has led to many Canadian sports success stories. But has it benefitted all, or has this vision failed our sports system at large? In broader terms, has our focus on owning the podium failed to provide Canadians with an appreciation of the intangible values of sport that go far beyond just podium finishes?

Today, as we make it our mandate to win Olympic medals, fewer and fewer young Canadians are staying in sports. In the complex system that is youth sport today, many individuals face barriers to participation from the heavy financial costs, extreme schedules, and pressure of early specialization. The narrative in high-performance sports becomes confused with a narrative of winning. At the Olympic level there is constant pressure to perform, because Sport Canada funding is closely tied to results that are reassessed annually. Those who fail to perform face very real consequences in the future of their sport's funding, and as a result others may never get the support they need. For example, after a poor performance at the Winter Olympic Games in Sochi 2014, the sport of skeleton was faced with a reduction of over $800,000 the following season. In the world of Canadian sport there is still a constant negotiation of trying to define the value of experience versus the value of the outcome.

An exploration into and valuation of failure is vital, because our current sport funding model does not accept failure. It does not give athletes and sports organizations the permission to fail because so much is tied in to their success. This fractures the sports system and the development of all Canadians. As Canadians' participation in organized sports continues to decline, who will be the next generation of champions? Who will rise up in the face of challenge and learn from failure? If we never recognize and value failure, how will we succeed?

My own success in Rio would not have been achieved had it not been for significant personal, emotional, and sporting failures. When we fail, we are given a powerful gift that enables us to understand ourselves and our true natures. We can use failure to see the other side of success and to shape our destinies toward a better future.

Failure is intertwined with the human experience. Because I have succeeded, I can talk about failing, because I have done both. I can look back and say the reason that I triumphed is *because* I failed. But life is often sloppy and incomplete. I have no mythical tale of triumph. No hero's journey. There has been failure along every single step of my journey, and sometimes a triumphant return, sometimes a gritty path back to another version of success. I have hit my breaking points at the most inopportune times, when I made mistakes or bad things happened. Failure is not the defining *moment*, it is the seconds, hours, days, even weeks around that moment that come to define who we are. Failure can be so fleeting, but equally so absolute, if we choose to let it be. Ultimately, with each failure we have a choice. We can choose to let it define us or we can choose to progress onwards and upwards.

CONCLUSION

Canada's Crossroads

Our country is at a crossroads. As our authors have shown, we are not an island unto ourselves. We are a part of a bigger global fabric. The world has shrunk and we are exposed, perhaps as never before. These are indeed tumultuous times, as the many examples of failure in this book have clearly illustrated; ours is not an easy world to live in. While the pessimists will think that this book has done nothing but vandalize our national pride and our Canadian identity, hopefully those of us who choose to look at our own personal failures, even if only for a little while, have seen how we have stumbled into an incredible opportunity as a country. As the world rages on all around us, Canada has remained strong through its accomplishments *and* its failures.

Of late, citizens from all around the world are choosing to be represented by governments with dangerously nationalistic under-tones. This new rise of nationalism is a cause for great concern, and should be for every Canadian. It creates fear and generates oppos-ition by creating an us-versus-them mentality — hopefully not felt at the same level within our country. The rise of new nationalist gov-ernments gives greater purpose to our nation of failed experiments. While, in many cases, the great social experiment that is Canada has failed in many ways, from the mistreatment of Indigenous communities, to a lack of empathy toward social issues such as homelessness, we remain, as a country, the best counterweight to a world showing its discontent through an extremist lens. Canada has not given way to the tendencies seen in other nations. We are choosing dialogue over fear, embrace over distance, and compassion over opposition. An open and honest conversation on failure is a must if we are to remain an effective counterbalance to the political events shaping our globe over recent years.

If the world's nationalist trends are not cause for concern, then perhaps the state of the world's economy should be. Again, Canada is not perfect in this domain. We have a growing amount of our national wealth in the hands of fewer and fewer Canadians, and our financial systems are not immune to global turmoil. Yet we choose to open our doors to trade. We choose not to threaten with walls but instead build a more prosperous North American continent *with* our neighbours. In this climate of economic uncertainty, Canada has choices to make. Will we be a nation promoting dialogue and global prosperity, or will we close our borders as other countries are threatening to do?

The pace of technological change is another major challenge we face. Information-processing power is growing exponentially, growing at such a rate that artificial intelligence is no longer a topic reserved purely for science fiction writers. Wearable technology is not a fad — soon we may all be walking computers. We humans have been harnessing resources for years in order to master our environment. Inuit created sunglasses. We Canadians invented special watches for train conductors, designed to withstand the shock and routine pounding experienced on our first railways. More recently, the Walkman revolutionized how we listened to music. We grow closer and closer to becoming cyborgs with every implant and every modification we make to ourselves. Thanks to technology, we are closer than ever before to being able to tackle some of the world's largest problems. But this technology can also destroy our world in ways we never thought imaginable. Is Canada a player in this new technologically advanced society? Do we have the wherewithal and the necessary entrepreneurs to drive our country to its next destination? Can we, as a nation, even participate in the transformation going on around us?

We should strive, at all costs, to remain relevant in this domain. It is imperative to our country because technological expertise will be required to deal with another major challenge, one that has the potential to destroy our planet. Climate change is real, and it is here. Whether or not it is human-made is irrelevant. For millennia, we humans have harvested our planet of all types of resources such as coal, food, water, and other materials and commodities needed for survival. It seemed that all we needed was at our fingertips. But with a population in perpetual growth, our old ways of managing our planet's resources are simply no longer adequate or acceptable. In this new reality, Canada has much to gain and lose from not being able to position itself on a global scale. Canada is blessed with many natural resources. However, have we been a responsible steward of them? Can we truly say we have the right ideology and concepts in place to be a positive example of responsible resources management to the world? We must ensure that we have our own house well in order before we try to answer this question.

All of these challenges and global struggles are bringing our country to a defining moment. There are important decisions to be made, and it is the next generation of Canadians who will be saddled with the burdens of these decisions. While the previous generations of Canadians can and should celebrate well-deserved successes, it is our younger Canadians, such as my own two children, who must look to the future using the lessons learned from our mistakes. This is imperative if we are to have a welcoming, just, and prosperous Canada for the next generations. That is why this book has come to be. Because it is important for tomorrow's Canadians to understand the world we are leaving to them. Canada is not perfect. Canada is beautiful and amazing, yet it is in a precarious position and faces many challenges. The only way to make sure that our country remains on the path we desire is to ensure we learn from our shortcomings and create the Canada of the future, not only for ourselves, but for the world. A baton is being passed to a new generation of Canadians. This new generation, like the generation before it, will learn to tailor the existing definition of what it means to be Canadian. The younger and older contributors to this book, men and women from various communities across the country, have all contributed to illustrating what it means to be Canadian through our failures, not just our successes.

A Culturally Sensitive Canada

As Professor David McNab illustrates in his chapter on Canada's Indigenous policies, we failed our First Nations long before 1867. Let us not write about how we are faced with an amazing opportunity to re-engage with Indigenous populations across this land. Instead, let us firmly and bluntly acknowledge the failures of the past. I say this not because I want to shame Canada as a nation, but rather because, as a Canadian who values inclusivity, I choose to accept multiple points of view. As we continue to speak of cultural acceptance, and we continue to accept more and more refugees from conflicts around the globe, we must not forget our first relationship, the most important one, that of our relations with Indigenous peoples. After all, we are all immigrants to their great land.

On a more individual level, Eric Chan's stories, outlining how his style of art was never truly accepted by the "establishment" until recently, are lessons in perseverance. He shows us how we must be more inclusive of the changing landscape around us. Digital technologies, globalization, and demographic changes all contribute to providing a fluid definition of what it means to be culturally inclusive. We are lucky to have artists such as Eric, who choose to fight the system and create what they feel is important to share with the world. I hope other Canadian artists and cultural enthusiasts come to realize that each voice is unique and needs to be communicated in a distinctive way.

Even the most accomplished artists among us, like Andy Nulman, the founder of the Just for Laughs Comedy Festival, have messed up along the way. His personal stories of failures could fill an entire book. Yet we see him as a symbol of entertainment success. His story is one of pushing boundaries further, always further, in order to make progress in the world of entertainment. Like authors David McNab and Eric Chan, he speaks of the impact of globalization on Canadian identity, and the need for us to assume our place in the world alongside household Canadian names such as Drake, Justin Bieber, and Alanis Morissette.

But what ties these three authors — Professor McNab, Eric Chan, and Andy Nulman — together as Canadians? One is a Métis professor from Manitoba, the other an Asian-Canadian Ottawa artist, and the third is a Jewish resident of Montreal. Absolutely nothing! That's what makes Canada the next global cultural powerhouse. Our country isn't uniformly vanilla; it is rich in diversity, and this quality is our strength. We have managed to ignore this strength by slotting our artists into neat little categories and by not rewarding avant-garde risk-taking. We need to look at ourselves in the mirror and admit, frankly, that it's a failure not to embrace our rich diversity boldly and openly. Once we have faced and accepted our differences and dealt with this failure and put it behind us, then we can become more creative and dynamic.

A Competitive Canada

I believe that our cultural failures can make us a global leader in the cultural space. But to be leaders, we have to be competitive. We have to want to compete *as a nation*. We can't say we're competitive just by pointing to competitive individuals within our great country. As Erica Wiebe, our treasured, gold-medal-winning Olympic wrestler shows us, winning is a result of being competitive. But winning is not everything, and it is not easy. We must redefine what we want to achieve in a way that sets goals meaningful to Canada, and not seek the goals that other countries have set for themselves. Canadian industries and companies compete in global markets every single day, and Canada needs to get tougher. We do not need to be like anyone else, or adopt another country's mantra. We need to be a Canadian style of tough: be respectful, but increasingly competitive. Too often we are dismissed with the labels of polite and charming. While those are not bad traits to have, I am sure Erica does not see them as representing her actions in the ring: "Introducing ... one of the world's most polite and charming wrestlers — Erica Wiebe!" There is a time for being polite and charming, and there is a time for hard, fair competition.

The same may be said for Canadian industry. We often wonder why our private-sector corporations cannot seem to pierce an apparent barrier between us and world. Nortel, BlackBerry, and the Avro Arrow haven't been able to achieve the levels of Microsoft, Apple, Boeing, or Lockheed Martin. Why is that? Is it because once again our polite demeanour gets in the way? I know the explanation is not that simple, and I do not pretend to have an answer. What I am stating, however, is that there is something in our waters that appears to prevent us achieving greatness in similar ways to our contemporaries like the U.K., the U.S., Japan, or even South Korea. Here again, I am sure our author Tom Jenkins, who serves as chair of both OpenText and the National Research Council Canada, would not want OpenText, the world's largest enterprise content management software company, to be known only as being "polite" in the global arena. Business may be polite, but it is, after all, competition.

The question arises: What will future generations of Canadians do in this setting? Will they continue to have an international reputation as honest brokers, as mere polite Canadians? Or can we, as a country, exercise Canadian competitiveness when it is required? As Dr. Amiee Chan details, there is no such thing as an overnight success in business. To be competitive, one must fail. Therefore, can we do better and take greater pride in our Olympic athletes, not only when the Olympics loom, but year-round? Can our industries be ruthless yet fair in global competition? I choose to believe our next generation of Canadians will find a way to strike a better balance than we have so far achieved, but only if they hear the stories of failure from our great competitors such as Amiee Chan, Tom Jenkins, and Erica Wiebe.

A Global Canada

In these turbulent times — in an epoch of change — "the world needs more Canada." Yes, we heard you, President Obama. Speaking as your little neighbour to the north it is not always easy to bring more Canada to the world. But we agree, it is possible, and we should do more. In fact, we have already begun. We have heard from our great astronaut Robert Thirsk, writing about how he prepared to face failure in one of the most dangerous environments known to our species: the vacuum of space. Thanks to women and men such as Robert, Canada has a place in space, and there is no grander destination for our species than space exploration. According to numerous experts, the very survival of our species may depend on it. But what is Canada's largest contribution to the space mandate? Without significant industry (in comparison to the U.K., the U.S., China, and Japan, and some other industrialized countries), what can Canada bring to the table? The answer could be as simple as: our values. We are the world's success story in social integration. We are understanding, we collaborate well with others, and we listen. While these traits may seem frivolous, as Robert illustrates quite clearly, these are character traits that are an absolute must for our species as we progress outward into space.

We can also lead the world in how we govern ourselves. Thomas Townsend illustrates that Canada is not perfect, nor are some of the professionals working in our public service. Yet they strive, every day, to make Canada a better place, at home and abroad. Canada has not been immune to its share of public scandals during the past 150 years. Nevertheless, we have learned from our public governance mistakes because we have assessed them and spoken openly about our short-comings. It is this type of dialogue that our nation can bring to the international community. We must return to the days of old when Canada was a diplomatic broker, a source of trust, and a promotor of dialogue. Our public service values are still very much present, even though the world may, at times, have forgotten us.

Furthermore, as our planet continues to shrink with transportation and technological innovations growing at an ever increasing pace, Canada can also lead in the field of health care. Our pioneers have led in global immunization. Our research centres, such as the National Microbiology Laboratory in Winnipeg, are among the best in the world. Is our health care system perfect? Absolutely not. We only need to go to the emergency room on a Friday evening to experience capacity issues in the form of waiting lists and endless queues. As Dr. Frank Plummer shares with us, our health care professionals are not immune to mistakes and failures. In fact, one could argue that our entire health system has achieved what it has because of failure. From insulin to the Ebola vaccine, breakthroughs only happen with failure. As a nation we must embrace this failure and not take a short-sighted approach to research, medical investments, or other facets of our national health system. This is not only to serve Canadians in Canada, but because we are leaders and can be even greater ones on the global stage. With a shrinking world comes increased health risk. We must continue to fail fast in order to better our place in the global fight against pandemic and disease.

Indeed, the world of tomorrow is an arena in which Canada must be present. We have much to offer. Our failures can help other nations to grow. With their growth, our nation grows as well. Not being shy in engaging in our shortcomings makes us stronger, better international partners for nations with good intentions. We can even teach others the art of humility, for admitting one's failures is the ultimate exercise in humility. An open dialogue on failure would no doubt also make this world a better place. The world needs more Canada; it needs Canada's failures in order to grow.

A Humble Canada

While it is true that we could be a Canada that is more culturally accepting, more competitive, and a more prominent global actor, we must nevertheless remain a humble nation. We must do so because we have failed on so many fronts. As Professor McNab shows us, our country has failed Indigenous peoples on too many levels. While we appear to want to fix this relationship, at least on the surface, we — the non-Indigenous population of this country — should not be determining the acceptable level of effort to repair centuries of abuse directed toward this land's Métis, First Nations, and Inuit populations.

Our nation has also failed our women. To be more precise, we have failed to encourage the contribution of women in the fields of science, technology, engineering, and mathematics. As Dr. Nausheen Sadiq illustrates, the struggle is real for our country's women to integrate into such fields and then advance within these domains. From salary inequality in our universities, to sexual harassment, to not providing the right environment for our girls to even consider entering such fields of study, we absolutely must do better nationally to assure gender parity in science. The very competitiveness of our country, our economy, and our social fabric depends on our ability to correct this wrong.

We are too imperfect to take an arrogant stand as a nation. Admitting our imperfection to ourselves is what can make our nation strong. To be humble, competitive, culturally sensitive, and globally in tune with our neighbours makes us Canadians. Accepting that our nation is as much defined by our failures as by our successes will help us achieve these objectives.

A Final Word

This book is about Canadian failures, and if you are reading this, at this very moment, you have made it through the entire process.

Congratulations! You are a proud Canadian.

Yet this book is not about you. In fact, it is also not about the authors who showcase their personal failures for all to see. No, this project is designed for the next generation of Canadians. This book is for our children's children, so they may not look at us and think that we, as Canadians, only celebrated stories of success. That we sat on our laurels contemplating our (apparent) amazingness. That we were so ignorant that we left these future Canadians with a planet worse than the one we inherited at the beginning of our own journeys.

As a result, the following is addressed to my unborn grandchildren and, in fact, to their children. It is also addressed to the unborn children of you who have read this far.

As the Canadians who went before you, I want to tell you that we were not perfect. We were not even amazing. As Canadians, however, we did try our best at everything we did. We wrestled our hearts out. We grew our businesses in the best manner we could. With little means we invented new medical breakthroughs. We laughed at ourselves in true Canadian fashion. Our art and our expressions changed with the times as we adapted and learned, at times painfully. Canada's governments attempted to do better than their predecessors because it was the right thing to do, even if what they achieved was not perfect. We began to acknowledge the damage we caused this land's Indigenous peoples, although it may be up to you to truly fix this relationship. And lastly, through science, we tried to advance humankind, with the help of both women and men, because we finally realized that science should be gender neutral. Science appreciates all contributions, regardless of their sources. And along the way, we made countless mistakes. But remember this: please treat each of our mistakes as you would a delicate flower. Our mistakes need to be nurtured, understood, and developed. They must be treated with respect, so they may help grow a better, stronger, more accepting, and just Canada.

Yes, we are a nation of failures. I for one am proud to admit it, because these failures make me who I am today, and hopefully help shape who you could be tomorrow: a cultured, understanding, yet fiercely competitive Canadian who understands the role you must play on this shrinking, sometimes troubled rock of ours. Because it is you who has the potential to take our failures and turn them into something amazing. Equipped with our failures, you can truly make the entire world a little bit more like Canada.

— Alex Benay

ACKNOWLEDGEMENTS

To write a book was not an easy decision to make.

I want to thank my wife for supporting me in this project and making sure I stood firm in what I believe. She pushed me to believe in my own abilities, despite my failures, and to ignore the critics. I love you, babe, and thank you!

I would also like to acknowledge my children, Cloee and Gabe, for their patience in dealing with me as I fumbled through this project. As any good parent is, I am often too hard on my children, yet I hope they can learn to appreciate this book as they, too, fumble and fail in life.

Mom and Dad: you picked me up off the ground every time I failed as a child, and I would not be who I am today, filled with failure and proud of it, without both of you!

Lastly, there is also one special person I would like to thank: David Sutin. During my time at the Canada Science and Technology Museums Corporation, David learned to take my enthusiasms and occasional downright craziness, and channel it into meaningful and cohesive thought. David is the reason this project was possible. He worked with all of the contributing authors, with me, and with a host of other characters to make this project a reality. David, you are nowhere near a failure, and I want to take a second to thank you as a friend, as you have helped me realize my dream of publishing a meaningful piece of work for Canada and the world.

— Alex Benay

IMAGE CREDITS

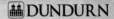